# Technological Trends in Water Sector for a Sustainable Solution

Sabarna Roy & Kaustav Ray Chaudhury

First published in 2021 by
BecomeShakespeare.com

One Point Six Technologies Pvt Ltd,
123, Building J2, Shram Seva Premises,
Wadala Truck Terminus,
Wadala (E), Mumbai - 400037
T:+91 8080226699



ISBN: 978-93-5458-067-3

## *Dedication*

Young Engineers at Workplaces

"There's nothing I believe in more strongly than getting young people interested in science and engineering, for a better tomorrow, for all humankind."

- Bill Nye

# Table of Contents

# 1

## KEY DEVELOPMENTS IN INDIAN IRRIGATION SECTOR

Sabarna Roy[1]

1. *Business Development, Applications Technology, Engineering and Strategy Department, Senior Vice President, Electrosteel Group, Kolkata-700017, India*

Kaustav Ray Chaudhury[2]

2. *Business Development, Applications Technology, Engineering and Strategy Department, Senior Executive, Electrosteel Group, Kolkata-700017, India*

**Abstract:** The key developments in the Indian Irrigation sector have been presented by briefly discussing the following topics:

1. Present Indian Irrigation Challenges and Way Forward.

2. Regulatory Interventions in Water Sector in India (Maharashtra's Example).

3. Roles and Activities of Maharashtra Water Resources Regulatory Authority (MWRRA).

4. Major Achievements of MWRRA.

5. Indian National Committee on Irrigation and Drainage (INCID).

6. Irrigation, Command Area Development and Micro Irrigation in India.

7. Flood Management and Its Measures.

8. Capacity Building.

**Keywords:** Irrigation, developments, MWRRA, INCID, CWC, water efficiency, flood forecasting, sustainability, Micro-Irrigation.

## 1. PRESENT INDIAN IRRIGATION CHALLENGES AND WAY FORWARD

It is known that 89% of water in India is consumed for irrigation. Today, the major concern is the lack of sustainability of water sector in India. The Indian Government has adopted the path for self-sufficiency in food grains through irrigation by creating a large number of dams and canals. India has extracted a colossal amount of groundwater which accumulates to 25% of the world's ground water storage, resulted in the construction of around 50 million water retaining structures [1]. The ways that should be adopted for water sustainability and water asset management are:

- ❖ *India needs to emphasize modernization of water systems to increase water efficiency.*

- ❖ *It is now becoming important to put some yardsticks to measure the efficiency of irrigation projects.*

- ❖ *The cropping pattern needs to be carefully selected based on the water availability of that region.*

- ❖ *Focus on Micro Irrigation should be increased.*

- ❖ *To channelize and utilize efficiently the rainwater which is available during the 3 months' of monsoon period for the remaining 9 months [1].*

- ❖ *India is the greatest extractor of ground water in the World (combined that of the USA and China). This is creating drying up of the aquifer. Ground Water Management programs are being taken up by the Government.*

- ❖ *New schemes like Atal Bhujal Yojana has been started to involve participatory groundwater management on the demand side so that there is a parity between supply-side and demand side.*

- ❖ *Finally – the emphasis is to be given by all the States as to how to irrigate more area with less water.*

## 2. REGULATORY INTERVENTIONS IN WATER SECTOR IN INDIA (MAHARASHTRA'S EXAMPLE)

Water Regulatory Authorities in India have resulted in State Water Policy in different states of Maharashtra, Uttar Pradesh and Jammu and Kashmir which has helped better to control the water usage.

Key features of the State Water Policy of Maharashtra were the incorporation of the following points:

- ❖ *Separate water tariffs were issued for the agricultural sector, domestic sector, and industrial sector.*

- ❖ *Implementation of water entitlement was another major inclusion.*

- ❖ *Integration of regulatory guidelines in three different levels, i.e., Policy level, Institutional level, and Administrative level.*

## 3. ROLES AND ACTIVITIES OF MAHARASHTRA WATER RESOURCE REGULATORY AUTHORITY (MWRRA)

Maharashtra Water Resource Regulatory Authority (MWRRA) was established in June 2005, enacted and functional since August 2006. The organisation has a Chairman and 4 expert members covering Water Resources Engineering, Law, Water Economy and Ground Water [2].

The roles and activities of the organisation which aims to provide sustainable water security to the State of Maharashtra are:

- ❖ *Regulating both surface water and groundwater of the state.*
- ❖ *To determine bulk water tariff for agriculture, industrial domestic and other purposes.*
- ❖ *To accord clearances to water resources projects.*
- ❖ *To act as the State Ground Water Authority.*
- ❖ *To support enhancement and preservation of water quality.*

## 4. MAJOR ACHIEVEMENTS OF MWRRA

The major achievements of the Maharashtra Water Resources Regulatory Authority (MWRRA) [5] are:

> *Regulatory interventions for sharing deficit in dry years by devising formulae and methodologies for judicious sharing among different sectors, and thus reducing the conflicts between various stakeholders.*

> *Regulatory interventions for promoting efficient water use and reducing wastage in the non-irrigation sector through proper tariff policies, budgeting and audit reports increases conservation awareness among users.*

> *Regulatory interventions for promoting efficient water use and reducing wastage in the agriculture sector through the promotion of micro-irrigation by tariff incentives and thereby increasing the area under micro-irrigation systems.*

> *Adjudication of water disputes by constituting transparent and trustworthy platforms for addressing grievances caused improvement of water governance in Maharashtra. MWRRA has disposed of total of 41 disputes in 14 years.*

> *Capacity building, knowledge sharing and motivation for action through collaboration with various governments on both national and international scale, academic institutions in conjunction with organising various conferences.*

**Although water reforms take place at a slow rate due to conflicts and domination of the socio-political system, MWRRA has done extensive work in bulk water tariff, equitable distribution of water in sub-basin and dispute resolutions. In its 14 years long journey the organisation has improved water governance in Maharashtra, reduced water conflicts and has enhanced the water use efficiency in agriculture significantly.**

## 5. INDIAN NATIONAL COMMITTEE ON IRRIGATION AND DRAINAGE (INCID)

❖ Central Water Commission (CWC) [3] is hosting the Indian National Committee on Irrigation and Drainage (INCID). INCID, CWC and ICID [4] closely cooperate in the water resources sector contributing towards knowledge dissemination: workshops, conferences, seminars etc.

❖ Ministry of Jal Shakti, in August 2019, re-constituted INCID as a dedicated national committee with representation from various states and union ministries, training and research institutes, NGOs, private sector, etc. for bringing ICID to the doorsteps of country-level development.

❖ INCID have developed several guidelines, manuals and books for consolidating knowledge in the irrigation management field.

In line with the objectives of ICID, CWC is engaged in various activities in the fields of:

❖ *Irrigation, development, and management.*

❖ *Development of hydropower energy.*

❖ *Flood Forecasting.*

❖ *Rehabilitation and modernisation of dams and irrigation projects.*

❖ *Trans-boundary and inter-state water management issues.*

❖ *Capacity Building.*

## 6. IRRIGATION, COMMAND AREA DEVELOPMENT AND MICRO IRRIGATION IN INDIA

- ❖ Irrigation Potential has increased from 22.6 Mha in (1951) to 126.73 Mha by the end of 2017.

- ❖ Food Grain Production increased from 51 MMT (1950-51) to 295 MMT (about 6 times) by the end of 2017.

- ❖ Accelerated Irrigation Benefit Program (AIBP) [8]: Initiative of Government of India for last-mile funding and technology support for early completion of irrigation projects Flood Forecasting [2].

- ❖ Command Area Development (CAD) Program was launched by the Government of India in 1974-75. Restructured and renamed as Command Area Development and Water Management (CADWM) in 2004. The objectives of CADWM Program are to bridge the gap between Irrigation Potential Created (IPC) and Irrigation Potential Utilized (IPU). It also improves Water-Use efficiency and increases agriculture productivity and production. It provides last-mile connectivity in the delivery of irrigation water and Participatory Irrigation Management (PIM).

- ❖ National Mission on Micro Irrigation (NMMI) under the Ministry of Agriculture (as part of National Mission on Sustainable Agriculture) [7] provides financial support to farmers. The potential for Micro Irrigation was 69.5 M ha and achieved till now is 10 M ha.

- ❖ Recent initiatives on improving irrigation efficiency involve:

  - ➢ *Mandatory implementation of micro irrigation in at least 10% of the Command Area in AIBP and CADWM Programmes.*

  - ➢ *Emphasis on Pipe Distribution Network in order to avoid Land Acquisition as well to avoid wastage of water. CWC has prepared a Guideline on Planning and Design of Pipe Irrigation regarding this.*

## 7. FLOOD MANAGEMENT AND ITS MEASURES

- ❖ Extensive Flood Management is taking place by increasing construction of Embankments, Channel/ Drainage Improvement, Flood Plain Zoning as per model bill of 1975- circulated by Ministry of irrigation [2].

- ❖ Increase in the number of Flood Forecasting Stations and Flood Forecasting Expansion (Fig. 1).

- ❖ Modernization in Flood Forecasting by collecting Automatic Real-time data and communication. Three days' advisory forecast based on rainfall-runoff modelling and collaboration with Google for inundation forecast (7 levels forecast stations- 11000 km$^2$), this is planned to increase to 3 times during this year.

- ❖ Establishment of Water Resources Information System (India-WRIS) for easy availability of water-related data for planning, development, and research.

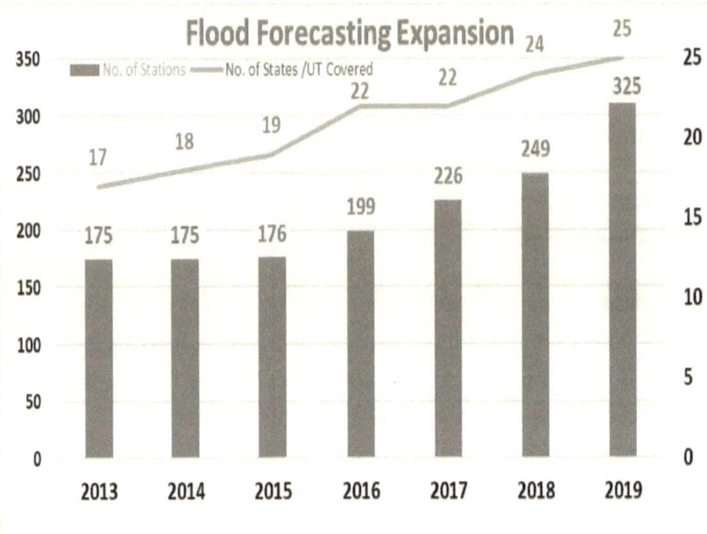


| State/UT | Level Forecast Station | Inflow Forecast Station | Total |
|---|---|---|---|
| Andhra Pradesh | 10 | 9 | 19 |
| Arunachal Pradesh | 3 | 0 | 3 |
| Assam | 30 | 0 | 30 |
| Bihar | 40 | 3 | 43 |
| Chhattisgarh | 1 | 2 | 3 |
| Gujarat | 6 | 7 | 13 |
| Haryana | 1 | 1 | 2 |
| Himachal Pradesh | 1 | 0 | 1 |
| Jammu & Kashmir | 3 | 0 | 3 |
| Jharkhand | 2 | 15 | 17 |
| Karnataka | 1 | 14 | 15 |
| Kerala | 3 | 2 | 5 |
| Madhya Pradesh | 2 | 9 | 11 |
| Maharashtra | 8 | 15 | 23 |
| Odisha | 12 | 7 | 19 |
| Rajasthan | 2 | 11 | 13 |
| Sikkim | 3 | 5 | 8 |
| Tamil Nadu | 4 | 11 | 15 |
| Telangana | 5 | 7 | 12 |
| Tripura | 2 | 0 | 2 |
| Uttar Pradesh | 39 | 4 | 43 |
| Uttarakhand | 4 | 2 | 6 |
| West Bengal | 12 | 4 | 16 |
| NCT Delhi | 2 | 0 | 2 |
| Daman & Diu | 1 | 0 | 1 |
| Total | 197 | 128 | 325 |

**Fig. 1 Flood Forecasting Expansion**

## 8. CAPACITY BUILDING

- ❖ In view of COVID, National Water Academy (NWA) [6] (training institute of CWC) and ICID have recently joined hands to use Digital Learning tools for trainings [2]:

- ❖ Principally it has been agreed to collaborate for conducting Training and Capacity building through Digital Learning in:

    - ➢ *Irrigation and Water management including Micro-irrigation.*

    - ➢ *Water use efficiency, Flood and Drainage.*

- ❖ It is under the process of development and training material is provided by NWA, and CWC is under review.

- ❖ Modular Object-Oriented Dynamic Learning Environment (MOODLE) platform available with ICID would be used.

- ❖ The Target Audience for the Distance Learning would be Water Resources and Irrigation professionals, farmers, NGOs, Agriculture extension staff, Students, International participants especially from the member countries of ICID etc.

## 9. CONCLUSION

Regulatory policies in India are changing towards a more sustainable environment with the implementation of water entitlement in the agricultural sector. Modernization in tools to account for water efficiency, data collection, storage and mapping and hands-on training of water resources personnel are to be executed. Regulations on bulk water tariff, equitable distribution of water in sub-basin and dispute resolutions are being handled on a serious note. The importance and mandatory implementation of Micro-Irrigation and Flood Forecasting with the help of innovative technologies will be a keystone in the forthcoming years.

## 10. REFERENCES

[1] Keynote Address by Shri U.P. Singh, Secretary, Ministry of Water Resources/Minister of Jal Shakti, Government of India on a webinar titled "*Agricultural Water Management Strategies in Changing Situation*" organised by International Commission on Irrigation and Drainage (ICID).

[2] Presentation on '*Regulatory Interventions in Water Sector in India – Maharashtra's Examples*' by Shri K.P. Bakshi, MWRRA Chairperson, Mumbai on a webinar titled "*Agricultural Water Management Strategies in Changing Situation*" organised by International Commission on Irrigation and Drainage (ICID).

[3] http://cwc.gov.in/.

[4] http://icid-ciid.org/.

[5] https://mwrra.org/.

[6] https://nwa.mah.nic.in/.

[7] National Mission on Micro Irrigation, Government of India, Ministry of Agriculture.

[8] http://mowr.gov.in/.

# 2

# DUCTILE IRON PIPES IN TRENCHLESS APPLICATION

Sabarna Roy[1]

1. *Business Development, Applications Technology, Engineering and Strategy Department, Senior Vice President, Electrosteel Group, Kolkata-700017, India*

Kaustav Ray Chaudhury[2]

2. *Business Development, Applications Technology, Engineering and Strategy Department, Senior Executive, Electrosteel Group, Kolkata-700017, India*

**Abstract:** Ductile Iron Pipe has been the pipe of choice when it comes to installing water and sewer pipelines because this material can withstand the toughest conditions. It can be used for installing new pipelines or for replacing existing pipelines. Two of the most important methods used in trenchless technology, namely horizontal directional drilling, and pipe bursting, are increasingly making use of ductile iron pipe because of its versatility and toughness. It is possible to assemble these pipes quickly in cartridge-style, making this type of pipe a great option for undertaking repair or installation work in places with heavy human and vehicular traffic. When space is an issue, it is also possible to pull or push these pipes through the ground using long strings. The objective of the paper is to discuss the following objects in connection with the use of Ductile Iron Pipes in Trenchless application:

- Trenchless Technology.
- Horizontal Directional Drilling (HDD).
- Pulling Force Resistance.
- Pulling Force Resistance Determination.
- Allowable Angular Deflection Minimum Values.
- Double Chambered Restrained Joint with Weld Bead.
- Dimension and Pulling Force Resistance (Double Chambered Restrained Joint with Weld Bead).
- Aggressive Ground Conditions.
- Pulling Heads.

**Key words:** HDD, PFA, PFR, DE, DI, OD

## 1. Introduction

The ductile iron pipe has been the pipe of choice when it comes to installing water and sewer pipelines because this material can withstand the toughest conditions. It can be used for installing new pipelines or for replacing existing pipelines. Two of the most important methods used in trenchless technology, namely horizontal directional drilling, and pipe bursting, are increasingly making use of ductile iron pipe because of its versatility and toughness. It is possible to assemble these pipes quickly in cartridge-style, making this type of pipe a great option for undertaking repair or installation work in places with heavy human and vehicular traffic. When space is an issue, it is also possible to pull or push these pipes through the ground using long strings.

Horizontal directional drilling (HDD), also known as directional boring, is a minimal impact trenchless construction technique used to install underground pipelines, conduits, cables, or any other utility along a pre-defined path.

## 2. Trenchless Technology

Trenchless technology, the application of a larger range of environmental protection areas can also be used. In a comparison of trenchless technology and conventional excavation pipeline construction technology, trenchless technology has certain advantages. Conventional excavation pipeline construction technology has the characteristics of simple operation, obvious effect, mature technology, and no technical risk. The state has a clear provision in the pipeline repair process requires a water supply operation to avoid waste of water and bring trouble to the construction. Therefore, the repair process with less time consuming, the quality of pipeline maintenance is better, the residents of the normal use of less water maintenance methods, the country is strongly supported. However, conventional excavation pipeline construction technology has a certain destructive; it will damage the original construction, affecting the traffic environment, destruction of natural structure, resulting in a lot of dust and serious air pollution. The use of conventional excavation pipeline construction technology maintenance section takes a long time to restore the original appearance.

## 2.1 Horizontal Directional Drilling (HDD)

General idea about HDD, and sequence of operations including method of HDD

(First step — Pilot bore, the Second step — Upsize bore, and the Third step — Pulling in).

**General:**

HDD is a steerable trenchless method of installing underground pipes along a prescribed bore path by using a surface-launched drilling rig, with minimum impact on the environment (see Figure 1). Directional boring is used where trenching or excavating is not practicable. It is suitable for a variety of soil conditions and projects, including road and river crossings.

The sequence of operations is generally divided into three successive steps.

*First step* — **Pilot bore**

The pilot bore is the first step in producing a bore, running from the starting point to the arrival pit, and is driven understeered control by a drilling head at the tip of a drilling string.

An aqueous suspension of bentonite emerges at high pressure from the drilling head, which

— drives the head forward,

— helps to cut the soil,

— carries away the cut material, and

— supports the bore.

The pilot bore is steered by controlled rotation of the drilling head; it is detected above the path of the bore by radio signals, gyroscope, or other means.

*Second step* — **Upsize bore**

During upsizing, an upsizing head is pulled through the bore, while rotating continuously; in this way, it enlarges the size of the pilot bore.

The soil that is cut away is carried out with the drilling mud, which also supports the bore.

The upsizing process is repeated with increasingly larger heads until the bore is of the desired diameter.

### Third step — Pulling in

For this third step, before pulling the pipes, three devices are attached to the drilling rods:

— a reaming tool;

— a rotary joint (which stops the string of pipes from turning with the reaming tool);

— a pulling head (connected to the string of pipes by mechanical locking).

As the pulling in progresses, drilling mud is pumped through the drilling linkage. It emerges from the reaming tool, carrying away the soil and reducing the frictional forces.

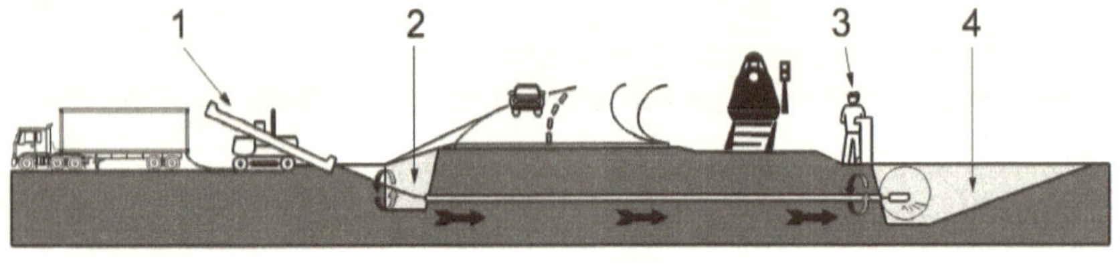


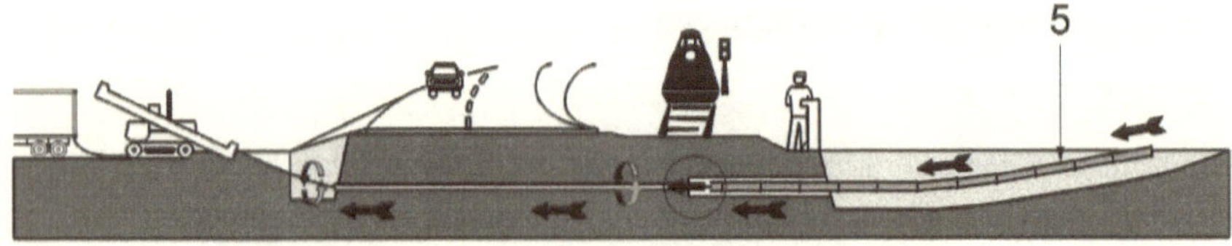


**Fig.1 Horizontal directional drilling method**

**Key**

1 drilling unit.

2 tunnel head earthworks.

3 piloting the drill.

4 tunnel exit earthworks.

5 anchored joint ductile iron pipes.

## 2.2 Pulling Methods

### 2.2.1 Assembly method

#### 2.2.1.1 Cartridge assembly method

The cartridge assembly option is defined as the assembling of individual sections of flexible restrained joint ductile iron pipe in a secured entry and assembly pit. The pipe sections are assembled individually and then progressively pulled into the bore path over a distance equivalent to a single pipe section. This assembly-pull process is repeated for each pipe length until the entire line is pulled through the bore path to the exit point.

#### 2.2.1.2 Ramp assembly method

The ramp assembly method is defined by the pre-assembly of multiple lengths of flexible restrained joint ductile iron pipe, with subsequent pulling installation into the bore path as a long pipe string. With this option, the contractor shall provide an entry ramp to the entrance of the bore path. The ramp shall be of sufficient length and grade such that anyone pipe joint does not exceed the allowable joint deflection at any point before the pipe string entering the bore path. The contractor shall be responsible for providing the necessary equipment or ground surface preparation to allow the pipe to be pulled back along the surface before the entry ramp and bore path without damaging the external coating/sleeving. The contractor shall repair any damage prior to the pipe section entering the bore path.

### 2.2.2 Drilling hole determination

The drilling hole diameter is technology and project-specific. The following are some basic guidelines:

a) HDD:

— for a straight project: the hole diameter less than 1.25 times pipe socket external diameter;

— for a curved project: the hole diameter less than 1.5 times pipe socket external diameter;

b) PB: a pipe that is larger than the pipe to be replaced may be used. The difference in diameters depends on the pipe material and the ground characteristics;

c) CM: a minimum gap of 50 mm should be used;

### 2.2.3 Pulling force control

The maximum pulling force used during the laying of the pipe shall be not higher than the pulling force resistance given by the manufacturer.

There are several methods to moderate the pulling force level. These methods include:

- the mass of the pipeline;

- friction to the ground, using appropriate fluids (characteristics of bentonite lubricant, adjustment, etc.);

- ballasting (with water and thicker pipes);

- reduction of the angle of deflection.

The contractor shall control the value of the pulling force by appropriate means. The pulling force progression shall be recorded accurately. A report, including the pulling force diagram, shall be made available by the contractor.

## 3. Pulling Force Resistance

For calculating pulling force resistance determination, pulling force minimum values, and allowable angular deflection minimum values should be taken into account.

The method for determination of pulling force resistance (PFR) shall be derived from the type test conducted on restraint joint to determine their pressure resistance, which is calculated as given by Formula (1):

$$PFR = \frac{PFA \times \pi \, (\emptyset DE)^2}{4 \times 10^4} \qquad (1)$$

Where,

- ➢ PFR is the pulling force resistance, in kilonewtons;

- ➢ PFA is the allowable operating pressure of the restrained joint, in bar;

- ➢ $\emptyset DE$ is the external diameter, as given in ISO 2531 or ISO 7186, in millimetres.

The safety factor for the PFA shall be at least 1.5 times PFA plus 5 bar.

The minimum value of the PFA of the restrained joint used in HDD or PB shall be 16 bar.

***NOTE 1:*** Higher pulling forces can be available; the pipe manufacturer can be consulted for their recommended maximum allowable pulling forces.

***NOTE 2:*** For a given restraint system, using a higher-pressure class to increase the pipe thickness can be a method to increase the pulling force resistance. If using an increase in pipe wall thickness, it is expected that care is taken to ensure the mechanical compatibility of the socket.

***NOTE 3:*** For long pulling lengths (e.g. 400 m), the allowable pulling force can be lower than the value calculated by Formula (1). In this case, it is intended that the manufacturer indicate this in the technical literature.

## 4. Allowable Angular Deflection Minimum Values

The minimum allowable deflection shall not be less than 3° for DN 80 to DN 300, 2° for DN 350 to DN 600 and 1° for DN 700 to DN 1200.

## 5. Double Chambered Restrained Joint with Weld Bead

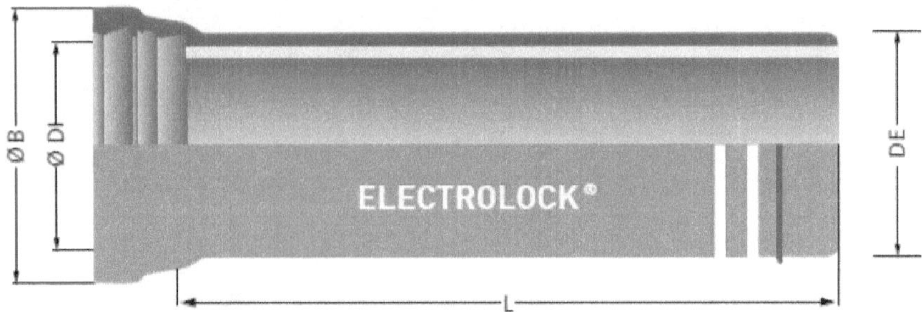


*Fig. 2*

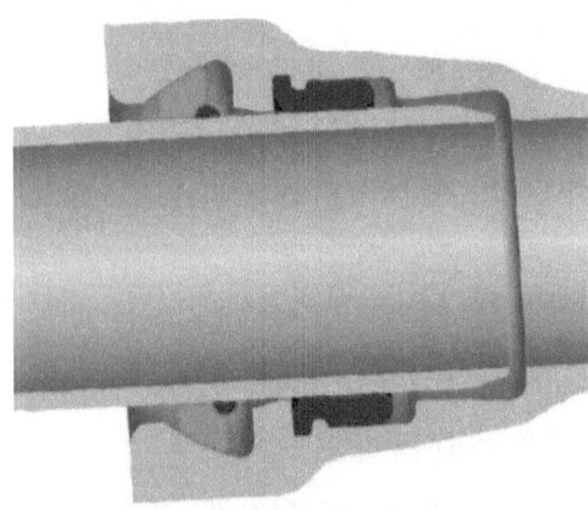

*Fig. 3*

Normal push-fit or mechanical joints in DI Pipes and Fittings do not provide significant restraint against longitudinal separation. Hydraulic thrust forces develop due to change of direction, reduction in diameter, and at the end of pipelines under pressure during operation. These forces may cause joint separation if anchor blocks or self-restraining anchoring joints are not provided in the pipeline.

## 6. Dimension and Pulling Force Resistance (Double Chambered Restrained Joint with Weld Bead).

| Pipe Size (mm) | Pipe Barrel OD (mm) | Pipe Socket ID (mm) | Allowable Deflection (Deg) | Minimum Curve Radius (Mtr) | PFA for standard application (Bar) | Allowable Pulling Load (KN) for Standard Application |
|---|---|---|---|---|---|---|
| 80 | 98 | 102 | 5 | 63 | 64 | 48 |
| 100 | 118 | 122 | 5 | 63 | 64 | 70 |
| 150 | 170 | 174 | 5 | 63 | 55 | 125 |
| 200 | 222 | 226 | 4 | 79 | 44 | 170 |
| 250 | 274 | 278 | 4 | 79 | 39 | 230 |
| 300 | 326 | 330 | 4 | 79 | 37 | 309 |
| 350 | 378 | 382 | 3 | 105 | 32 | 359 |
| 400 | 429 | 434 | 3 | 105 | 30 | 433 |
| 450 | 480 | 485 | 3 | 105 | 30 | 543 |
| 500 | 532 | 537 | 3 | 105 | 30 | 667 |
| 600 | 635 | 640 | 3 | 105 | 30 | 950 |
| 700 | 738 | 744 | 3 | 105 | 30 | 1283 |
| 800 | 842 | 848 | 2 | 158 | 25 | 1391 |
| 900 | 945 | 952 | 2 | 158 | 25 | 1753 |
| 1000 | 1048 | 1055 | 2 | 158 | 25 | 2155 |

*Table No.1: Dimension and pulling force resistance (Double Chambered Restrained Joint with Weld Bead)*

## 7. Aggressive Ground Conditions

Electrosteel recommends that pipe used for HDD to have additional external protection (tape wrap) if the project area has a history of extremely corrosive soil, or a soil survey determines the soil to be extremely corrosive.

## 8. Pulling Heads

*Fig. 4*

Electrosteel offers specially designed pulling heads for HDD application. These pulling heads can also be used for pressure testing the line.

## 9. Conclusion

The use of trenchless technology in the/a piping project can help save costs, protect the environment, and minimize surface disruptions. When it comes to Trenchless Technology, not having to dig a trench is the whole point. Trenchless technology is tunneling below the surface to install service lines like water or gas pipes, electric or telecommunication cables, without anyone noticing on the surface. This method also makes it possible to install utilities under rivers canals and other obstacles with no disruption of flow and with minimum or no damage to the environment. Trenchless methods have been used for the last 50 years, but these days, with advancements in technology for excavation and guidance systems, trenchless methods are becoming more commonly used around the world when soil conditions and location permits.

# 3

# POLYURETHANE COATING AND LINING ON DUCTILE IRON PIPES

Sabarna Roy[1]

1. *Business Development, Applications Technology, Engineering and Strategy Department, Senior Vice President, Electrosteel Group, Kolkata-700017, India*

Kaustav Ray Chaudhury[2]

2. *Business Development, Applications Technology, Engineering and Strategy Department, Senior Executive, Electrosteel Group, Kolkata-700017, India*

**Abstract:** In this Technical Paper we will discuss Polyurethane Linings and Coatings to Ductile Iron Pipe for transporting corrosive and abrasive fluids in a sub soil, that is extremely corrosive in nature. The capabilities of Electrosteel Group (ECL) in this regard will be discussed.

**Key words:** Polyurethane, lining, coating, ECL, ISI, test certificates, performance tests, provision, specification.

## 1. POLYURETHANE LININGS AND COATINGS

Two types of "special lining and coatings" process are implemented at Electrosteel Castings Limited (ECL) Bansberia Works [1]:

A polyurethane epoxy coating (heavy-duty) is applied to the outside wall of the pipe with a polyurethane lining (heavy-duty) inside.

❖ *ECL-Bansberia plant was built solely for the purpose of applying "special" types of coatings to ductile iron pips manufactured at the ECL-Khardah Works. The "special" coatings are designed to give greater external corrosion protection for pipes to be installed in aggressive soil conditions or for greater internal corrosion protection for pipes transporting highly corrosive liquids.*

* *The pipes which undergo the "special" external coating process are delivered with cement mortar lining already having been applied at the Khardah Works but with the external surface of the pipe uncoated (i.e. no zinc or other external coating on the pipe). The "special" external coating is applied in accordance with BS EN 15189.*

* *The pipes which undergo the "special" internal lining process are delivered with external zinc or zinc/aluminium coating having previously been applied at the Khardah Works. The pipes are then internally lined with the "special" standard polyurethane lining. The "special" Internal Lining is applied in accordance with BS EN 15655 - 2018.*

* *The bare pipes which undergo both "special" internal lining and external coating are hydrostatically tested at the Khardah Works before being transported to the Bansberia Plant for the "special" coatings and linings.*

## 2. PROVISIONS FOR POLYURETHANE EXTERNAL COATING AS PER BSEN15189-2006

BS EN15189-2006 – This document defines the requirements and test methods applicable to factory-applied external polyurethane coating (heavy duty) corrosion protection of ductile iron pipes and fittings conforming to EN 545, EN 598 and EN 969 [2].

* **STANDARD**: *BS EN 15189:2006 (for PU).*
* **BRAND:** *PROTEGOL UR COATING 32-49 (for PU).*
* MANUFACTURER: *TIB Chemicals, Germany (for PU).*

## 3. TESTING SAMPLE FOR PERFORMANCE TESTS

**Table 1 Testing Sample for Performance Tests.**

| Sl. No. | Test | EN 15189 Ref. Clause | Test Method Ref. Clause of EN 15189 | Sample |
|---|---|---|---|---|
| I | Chemical resistance | 6.1 | 7.2.1 | Detached Film |
| II | Impact Strength | 6.2 | 7.2.2 | Coated pipe sample |
| III | Indentation Resistance | 6.3 | 7.2.3 | Coated pipe sample |
| IV | Elongation at Break | 6.4 | 7.2.4 | Detached Film |
| V | Specific Coating Resistance | 6.5 | 7.2.5 | Coated pipe sample |
| VI | Ratio of Coating resistance | 6.5 | 7.2.5 | Coated pipe sample |

Sample of the following dimensions are cut from coated pipe for performing:

A) Impact Strength      600 mm X 150 mm

B) Indentation resistance test      100 mm X 150 mm

C) Specific coating resistance      175 mm X 200 mm

Detached PU film is collected from a polymeric sheet of 100 mm X 500 mm attached to the pipe body during coating for performing Chemical Resistance and Elongation at the break.

## 4. PERFORMANCE TESTS OF PU COATING

Performance tests conducted at Electrosteel Castings Limited, Khardah. Works for performance requirements of EN 15189:2006.

**Table 2 Performance Tests of Pu Coating.**

| SI No | Test | Sample | Date of Test | Duration | Result |
|---|---|---|---|---|---|
| 1 | Chemical resistance | Detached Film | 15.10.2013 | 200 days | Pass |
| 2 | Impact Strength | Coated pipe sample | 05.02.2013 | - | Pass |
| 3 | Indentation Resistance | Coated pipe sample | 30.01.2013 | 2 days | Pass |
| 4 | Elongation at Break | Detached Film | 05.02.2013 | - | Pass |
| 5 | Specific Coating Resistance | Coated pipe sample | 05.02.2013 | 100 days | Pass |
| 6 | Ratio of Coating resistance | Coated pipe sample | 05.02.2013 | - | Pass |

**Table 3 Performance Tests of Pu Coating with Requisite Clauses.**

| Sl. No. | Parameter | Requirement | Clause | Test method | Clause |
|---|---|---|---|---|---|
| 1 | Chemical resistance | Less than 15% weight increase after immersion Less than 2% weight loss after drying | 6.1 | Immersion in deionised water EN ISO 62 method 2 | 7.2.1.1 |
| | | Less than 10% weight increase after immersion Less than 4% weight loss after drying | | Immersion in diluted sulphuric acid 10% EN ISO 62 method 2 | 7.2.1.2 |
| 2 | Impact strength | 8 J/mm PU-coated pipe barrel 5 J/mm EP-coated spigot end (see EN 14901) | 6.2 | Dropping weight High voltage test | 7.2.2 |
| 3 | Indentation resistance | < 10% at 10 Mpa | 6.3 | Indentation test | 7.2.3 |
| 4 | Elongation at break | > 2.5% | 6.4 | Tensile test | 7.2.4 |
| 5 | Specific coating resistance in 0.1 M NaCl | $> 10^8 \ \Omega m^2$ | 6.5 | Resostovotu test towel method or vessel method | 7.2.5 |
| 6 | Ratio of coating resistance | > 0.8 | 6.5 | Res. 100 d/res. 70d | 7.2.5 |

## 5. ROUTINE TESTS OF PU COATING

Routine tests were conducted at ECL, Khardah Works for performance requirements of EN 15189:2006.

**Table 4 Routine tests of PU coating.**

| Sl. No. | Parameters | Requirements | Clause | Tests | Frequency | Clause |
|---|---|---|---|---|---|---|
| 1 | Surface preparation | SA 2.5 of EN ISO 8501-1 | 5.1 | Visual | 100% | 7.1.1 |
| 2 | Surface roughness | Ra > 10 $\mu$m | 5.1 | EN ISO 8503-1 | min. 1/shift | 7.1.1 |
| 3 | Appearance and continuity | Uniform and smooth | 5.2.1 | Visual | 100% | 7.1.2 |
| 4 | Minimum coating thickness | $(x - 2\sigma) > 700$ microns | 5.2.2 | Non-destructive instruments error ± 10% | min. 1/shift | 7.1.3 |
| 5 | Pipe ends painted parts | Length depending on type of socket | 5.3 | Appropriate measures | 10% | 7.1.4 |
| 6 | Repairs | Manufacturer's written instructions | 5.4 | High voltage test | 100% | 7.1.5 |
| 7 | Marking | Legible and durable | 5.5 | Visual | 10% | 7.1.6 |
| 8 | Non-porosity | No electrical break through at required test voltage | 5.6 | High voltage test instrument | 1 per 1000 pipes | 7.1.7 |
| 9 | Hardness | > 70 Shore D | 5.7 | Hardness test | min. 1/shift | 7.1.8 |
| 10 | Adhesion | > 8 Mpa at $23^0$C | 5.8 | Punch separation method acc. EN ISO 4624 | 1 per 1000 pipes | 7.1.9 |

## 6. TYPE TEST CERTIFICATES OF POLYURETHANE COATED DI PIPES

**Type Test Certificates for Polyurethane Lining certified by NSF is attached below [Fig.1]**

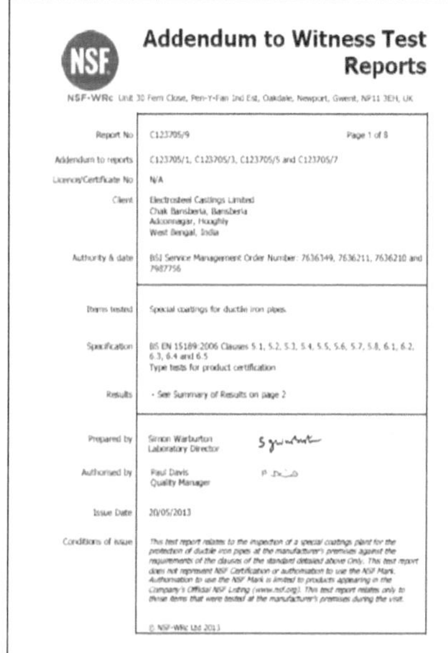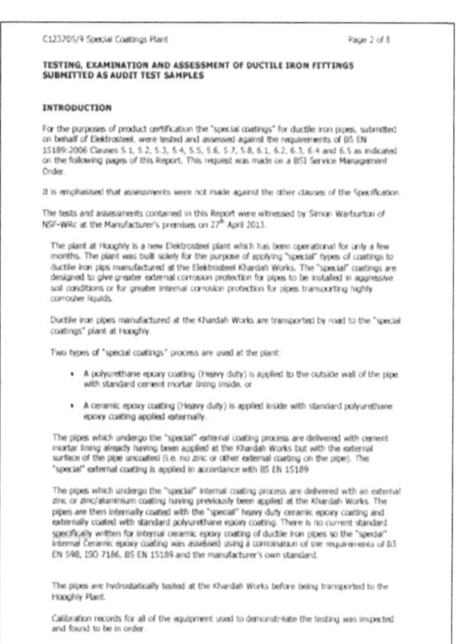

**Fig. 1 Type Test Certificates of Polyurethane Coated DI Pipes [3].**

## 7. SPECIFICATIONS FOR THE EIL PROJECT ON EFFLUENT DISCHARGE PIPELINE

It may be noted that the soil type along the pipe route is generally Silty Clay/Clay/ Silty Sand/ Sand/Silt. The chemical characteristics for Soil and Ground water are ranging as follows [4];

### Soil:

- *Sulphate ($SO_3^-$) - Nil to 95 (mg/l)*

- *Chloride - 164-506 mg/l*

- *pH - 7 to 9*

### Ground Water:

- *Sulphate ($SO_3^-$) - Nil to 95 (mg/l)*

- *Chloride - 108-347 mg/l*

- *pH - 7 to 9*

**The use of Polyurethane Coating as per BS EN 15189:2006 has been stipulated having a coating of 1000 micron.**

## 8. PROVISIONS FOR POLYURETHANE INTERNAL LINING AS PER BSEN15655-1:2018

BS EN15655-1:2018 – This document defines the requirements and test methods applicable to factory-applied internal polyurethane heavy-duty corrosion protection of ductile iron pipes and fittings conforming to EN 545, EN 598 and EN 969 [5].

❖ **Chemical resistance (Clause 6.1)**

The chemical resistance is determined by the change in weight of the polyurethane lining in neutral or acid conditions. When tested in accordance with 7.2.3 the weight increase respectively, weight loss shall meet the requirements given in Table 3 when compared to the original weight.

| Property | Unit | Test method | Clause | Requirement |
|---|---|---|---|---|
| Mass change after 100 days in deionised water at 50 °C | % | Immersion test | 7.2.3.2 | Less than 15 % weight increase |
| Mass change after subsequent drying | % | EN ISO 62, method 3 | | Less than 2 % weight loss |
| Mass change after 100 days in sulphuric acid 10 % at 50 °C | % | Immersion test | 7.2.3.3 | Less than 10 % weight increase |
| Mass change after subsequent drying | % | EN ISO 62, method 3 | | Less than 4 % weight loss |

❖ **Indirect Impact strength (Clause 6.2)**

*Due to handling activities, the PU-lined pipes may fall or get impacts from outside with minor plastic deformations which can cause damages on the lining. The minimum impact strength shall be determined in accordance with the test method defined in 7.2.4 with an impact energy E of at least 50 J. The lining shall subsequently show no damage when tested in accordance with 7.1.8. Water use efficiency, Flood and Drainage.*

❖ **Resistance to ovalization (Clause No. 6.3)**

*The requirements of EN 545 or EN 598 shall apply.*

❖ **Elongation at break (Clause No. 6.3)**

*The elongation at break shall be assessed by testing in accordance with the test method defined in 7.2.6. The lining shall have a minimum elongation at break of 2,5 %.*

❖ **Glass transition temperature (Clause No. 6.3)**

*The lining material shall conform to the limits of change in glass transition temperature ($\Delta T_g$) determined by Differential Scanning Calorimetry (DSC).*

❖ **Specific electrical resistance of the lining (Clause No. 6.3)**

*The specific lining resistance of the polyurethane lining shall be assessed by testing in accordance with the test method defined in 7.2.7.*

*The specific lining resistance of the polyurethane lining after immersion in a 0,1 M NaCl solution for 100 days shall be at least 108 Ω m2. The resistance after 100 days shall not be less than 80 % of the value after 70 days if the surface resistance of the lining after 100 days is only one decimal power above the minimum permissible value for 100 days. The test shall be carried out at room temperature (23 ± 2) °C.*

❖ **Abrasion resistance (only for waste water application) (Clause No. 6.7)**

*When tested in accordance with 7.2.8, the pipes shall not have an abrasion depth greater than 0,2 mm after 100 000 movements (50 000 cycles).*

**Note:** *In order to test the abrasion resistance of fittings, straight fittings as flanged pipes, etc. may be lined as fittings and tested according to 7.2.7.*

❖ **Materials in contact with water intended for human consumption (Clause No. 6.8)**

*When used under the conditions for which they are designed, in permanent or in temporary contact with water intended for human consumption, the polyurethane lining applied on ductile iron pipes and fittings shall not change the quality of that water to such an extent that it fails to comply with the requirements of national regulations.*

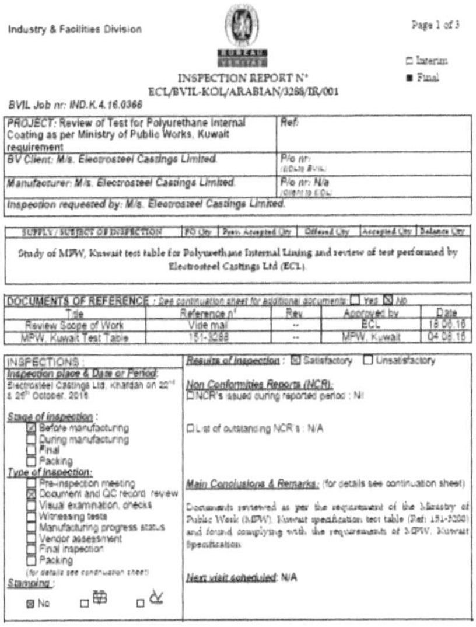

*For this purpose, reference shall be made to the relevant national regulations and standards, transposing EN standards when available, dealing with the influence of materials on water quality and to the requirements for external systems and components as given in EN 805.*

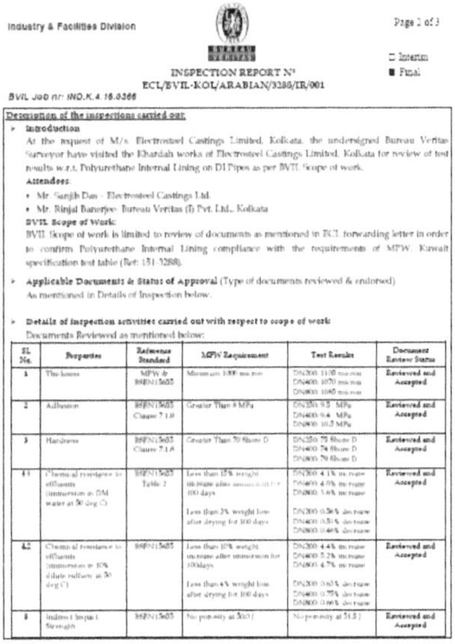

# 9. TYPE TEST CERTIFICATES OF POLYURETHANE LINED DI PIPES

**Fig. 1 9. Type Test Certificates of Polyurethane Lined DI Pipes [6].**

## 10. RESULT OF TYPE TEST FOR POLYURETHANE INTERNAL COATING

> Details of inspection activities carried out with respect to scope of work

Documents Reviewed as mentioned below:

| SL No. | Properties | Reference Standard | MPW Requirement | Test Results | Document Review Status |
|---|---|---|---|---|---|
| 1 | Thickness | MPW & BSEN15655 | Minimum 1000 micron | DN200: 1105 micron<br>DN400: 1070 micron<br>DN800: 1085 micron | Reviewed and Accepted |
| 2 | Adhesion | BSEN15655 Clause 7.1.9 | Greater Than 8 MPa | DN250: 9.5 MPa<br>DN400: 9.4 MPa<br>DN800: 10.5 MPa | Reviewed and Accepted |
| 3 | Hardness | BSEN15655 Clause 7.1.8 | Greater Than 70 Shore D | DN250: 75 Shore D<br>DN400: 74 Shore D<br>DN800: 79 Shore D | Reviewed and Accepted |
| 4.1 | Chemical resistance to effluents (immersion in DM water at 50 deg C) | BSEN15655 Table 2 | Less than 15% weight increase after immersion for 100 days<br><br>Less than 2% weight loss after drying for 100 days | DN200: 4.1% increase<br>DN400: 4.0% increase<br>DN800: 3.6% increase<br><br>DN200: 0.56% decrease<br>DN400: 0.51% decrease<br>DN800: 0.46% decrease | Reviewed and Accepted |
| 4.2 | Chemical resistance to effluents (immersion in 10% dilute sulfuric at 50 deg C) | BSEN15655 | Less than 10% weight increase after immersion for 100days<br><br>Less than 4% weight loss after drying for 100 days | DN200: 4.4% increase<br>DN400: 5.2% increase<br>DN800: 4.7% increase<br><br>DN200: 0.63% decrease<br>DN400: 0.75% decrease<br>DN800: 0.66% decrease | Reviewed and Accepted |
| 5 | Indirect Impact Strength | BSEN15655 | No porosity at 50.0 J | No porosity at 51.5 J | Reviewed and Accepted |
| 6 | Ovalization resistance | BSEN15655 | No damage at minimum 4 % Ovalization for DN200<br>6 % Ovalization for DN400<br>8 % Ovalization for DN800 | No damage at<br>4.8 % for DN200<br>6.4 % for DN400<br>9.0 % for DN800 | Reviewed and Accepted |
| 7 | Elongation at Break | BSEN15655 | > 2.5% | 4.2% for DN200<br>5.2 % for DN400<br>4.1 % for DN800 | Reviewed and Accepted |
| 8 | Abrasion resistance (50000 cycles) | BSEN15655 & BSEN598 | < 0.20mm | 0.09 mm | Reviewed and Accepted |
| 9 | Light ageing resistance (Outside Storage for 6 months) | BSEN15655 | Adhesion (>8 MPa) | DN250: 9.3 MPa<br>DN400: 8.9 MPa<br>DN800: 9.3 MPa | Reviewed and Accepted |
| 10 | Resin Content of PU material | MPW | 85% minimum | More than 85% | Manufacturer Confirmation was noted |

**Table 5 - Result of Type Test for Polyurethane Internal Coating.**

## 11. SPECIFICATIONS FOR THE EIL PROJECT ON EFFLUENT DISCHARGE PIPELINE

**Table 6 Effluent Characteristics from Ankleshwar facility [4].**

**As per the provided effluent characteristics, the use of Polyurethane lining with 1500 micron thickness has been stipulated.**

| S. No. | PARAMETER | UNIT | Value |
|---|---|---|---|
| 1 | pH | mg/l | 6.00 to 9.00 |
| 2 | COD | mg/l | 500 |
| 3 | BOD $_{3\,Days,\,27°C}$ | mg/l | 100 |
| 4 | Total Dissolved Solids | mg/l | ~ 10000 |
| 5 | Total Suspended Solids | mg/l | 100 |
| 6 | Sulphides, as S | mg/l | 5 |
| 7 | Phenolic Compounds (as C6H5OH) | mg/l | 5 |
| 8 | Oil and Grease | mg/l | 10 |
| 9 | Total Residual Chlorine | mg/l | 1 |
| 10 | Fluoride | mg/l | 15 |
| 11 | Free Ammonia | mg/l | ~ |
| 12 | Nitrate Nitrogen | mg/l | 50 |
| 13 | Ammonical Nitrogen | mg/l | 50 |
| 14 | TKN | mg/l | 50 |
| 15 | Vanadium | mg/l | 0.2 |
| 16 | Selenium | mg/l | 0.05 |
| 17 | Iron | mg/l | 3 |
| 18 | Copper | mg/l | 3 |
| 19 | Zinc | mg/l | 15 |
| 20 | Chromium 6+ | mg/l | 0.1 |
| 21 | Lead | mg/l | 0.1 |
| 22 | Cadmium | mg/l | 0.05 |
| 23 | Temperature | °C | Not more than 5°C above ambient water |
| 24 | Arsenic | mg/l | 0.2 |
| 25 | Mercury | mg/l | 0.01 |
| 26 | Manganese | mg/l | 2 |
| 27 | Nickel | mg/l | 3 |

## 12. ISI MARKING OF DI PIPES

The Indian Standards Institution (ISI) marking of DI pipes is mandated by the order of the Ministry of Commerce and Industry (Department of Industrial Policy and Promotion) dated 25th June 2009.

So, even if a customer does not ask for ISI marked DI pipes, there is no way but the DI pipe manufacturers are bound to supply ISI marked DI pipes as per this order.

In the said Gazette in Sl. No. 3 under the heading "Prohibition regarding the manufacture, storage, sale and distribution etc." it is written that [7] –

**Quote:**

*(1) No person shall by himself or through any person on his behalf manufacture or store for sale, sell or distribute Ductile Iron Pressure Pipes and Fittings which do not conform to the specified standard and do not bear Standard Mark of the Bureau on obtaining certification marks license:*

> *Provided that nothing in this Order shall apply in relation to the export of Ductile Iron Pressure Pipes and Fittings meant for export, which conforms to any specification required by the foreign buyer and such specification shall not, in any case, be less than the specified standard.*

*(2) The sub-standard or defective Ductile Iron Pressure Pipes and Fittings, which do not conform to the specified standard shall be deformed by the manufacturer beyond use and disposed off as scrap within three months.*

**: Unquote**

## 13. CONCLUSIONS

Polyurethane Lining should be adopted for corrosive and abrasive fluids and the thickness should be determined based on fluid properties.

Polyurethane Coating should be adopted for corrosive sub-soil and the thickness should be determined based on sub-soil properties.

## 14. References

[1] https://www.electrosteel.com/

[2] BS EN15189-2006 - Ductile Iron Pipes, fittings and accessories – External polyurethane coating for pipes - Requirements and test methods.

[3] https://www.wrcplc.co.uk/wrc-nsf.aspx.

[4] https://engineersindia.com/.

[5] BS EN 15655-1:2018 - Ductile iron pipes, fittings and accessories -Requirements and test methods for organic linings of ductile iron pipes and fittings.

[6] https://www.bureauveritas.co.in/.

[7] https://bis.gov.in/.

# 4

# RAINWATER HARVESTING, LEGISLATION AND MANDATES ON IT IN DIFFERENT STATES OF INDIA

Sabarna Roy[1]

1. *Business Development, Applications Technology, Engineering and Strategy Department, Senior Vice President, Electrosteel Group, Kolkata-700017, India*

Kaustav Ray Chaudhury[2]

2. *Business Development, Applications Technology, Engineering and Strategy Department, Senior Executive, Electrosteel Group, Kolkata-700017, India*

## ABSTRACT

Today, the primary concern is the lack of sustainability of water in India. India has extracted a colossal amount of groundwater, which accumulates to 25% of the World's Ground Water storage, resulted in the construction of around 50 million water retaining structures. Also, India is the greatest extractor of Ground Water in the World (combined that of USA and China). This is creating drying up of the aquifer. So, the Government is taking up Ground Water Management programs to emphasize the modernization of water systems to increase water efficiency.

So, having an alternative rainwater harvesting system would be very useful to the surroundings during emergency and non-emergency situations, especially to channelize and efficiently utilize the rainwater available during the three-month monsoon period in the remaining nine months.

This paper represents an attempt to detail out the characteristics and parameters of rainwater harvesting. Emphasis has been given to the operation and maintenance of such a system.

Many State Governments and Central Government Undertakings in India have already acted to ensure rainwater harvesting is compulsory. For example, West Bengal, Chandigarh, and Himachal Pradesh have made rainwater harvesting mandatory. And, cities like Mumbai and Bengaluru have also passed specific laws to enforce rainwater harvesting. In this paper, the various legislation and mandates adopted in India's states have also been discussed.

It is also notable to mention that NTPC set out a guideline in 2018 for making Rain Water Harvesting a more predominant system.

## <u>INTRODUCTION</u>

Water is an essential natural resource and it is the necessity of our lives. We use water for drinking, irrigation, industry, domestic use, transportation, and hydroelectricity production. Water demand is continuously increasing in developing India as the population increases. It requires overcoming the deficiency of surface water to meet our needs. To overcome this enormous water demand, rainwater harvesting (RWH) is an effective and economical tool for growing India. Rainwater harvesting is sustainable and economic development.

Harvesting of rainwater is the process of collecting rainwater from various surfaces such as a rooftop, sloped roof, roof tiles, or ground and store it for future purposes. In areas where there is excess rainfall, the surplus rainwater can be used to recharge groundwater through artificial recharge techniques. RWH can elevate the groundwater table by artificially recharging groundwater resources like tube wells, bore wells, and open wells.

### <u>Reasons Behind Rainwater Harvesting:</u>

1. Representatives of the Federal Government involved with the water supply planning process have determined that the present amounts of the available surface and groundwater supplies will not meet future water demand. Water conservation and the development of alternative water supplies is necessary to meet our growing demand for freshwater.

2. Rainwater harvesting is an alternative water supply approach that captures, diverts, and stores rainwater for later use and is available to anyone. Captured rainwater is often used as a potable water source. Another widespread use is for attracting and providing water for wildlife, pets, and livestock. Rainwater is also used for landscaping because the water is free of salts and other harmful minerals. Rainwater does not have to be treated with chemicals that have residual influences for most non-potable uses.

3. Implementing rainwater harvesting techniques directly benefits our

country and state by reducing the demand on the municipal and public water supply and reducing runoff, erosion, and contamination of surface water.

4.  Rainwater harvesting can also help prevent flooding and erosion, turning stormwater problems into water supply assets by slowing runoff and allowing it to soak into the ground. This also helps decrease surface water contamination with sediments, fertilizers, and pesticides in rainfall runoff.

5.  Rainwater harvesting can be used in small-scale residential landscapes and large-scale landscapes, such as parks, schools, commercial sites, parking lots, and apartment complexes. It can also be used in homes for commodes, and clothes washing etc.

## ADVANTAGES OF USING RAINWATER

1.  Promotes Conservation of Water: Rainwater harvesting promotes self-sufficiency and appreciation for water as a resource. It also encourages water conservation while providing an alternative water source.

2.  Conserves Energy: Because the centralized water system is bypassed through rainwater harvesting, this system will save energy. Many systems require only a small pump to create water pressure in household pipes, and many non-potable systems operate with gravity as the only driving force.

3.  Reduces Undesired Storm Water Runoff: Local erosion and flooding from impervious cover associated with buildings are lessened as a portion of the local rainfall is diverted into collection tanks, leaving less storm water to manage.

4.  Improves the quality of Water Supply: Rainwater is one of the purest sources of water available. Its quality almost always exceeds ground or surface water because it does not contact soil or rocks to dissolve minerals and salts. Also, captured rainwater will not come into contact with many pollutants which are often discharged into local surface waters or contaminated groundwater supplies.

5.  Supplies Nutrients to Plants: Rainwater often comprises nitrogen content that provides a slight fertilizing effect.

6.  Provides Naturally-Soft Water: Rainwater is considered soft water, and therefore significantly lowers the quantity of detergents and soaps needed for cleaning. Soap scum and hardness deposits do not occur. Water softeners are not necessary with rainwater as it is often with well water. Also, water heaters and pipes are free of the deposits caused by hard water and should last longer, thus saving money.

## USAGE OF HARVESTED RAINWATER

There are several usages of Rainwater harvesting in Landscape Watering, Wildlife Watering, Feeding Livestock and Pets, Groundwater Recharge, Reduce Storm water Runoff, Home application, Commercial and Industrial, etc.

Broadly there are two ways of harvesting rainwater:

i.   Surface runoff harvesting

ii.  Rooftop rainwater harvesting

**Surface Runoff Harvesting:** In urban areas, rainwater flows away as surface runoff. This runoff could be caught and used for recharging aquifers by adopting appropriate methods.

**Rooftop Rainwater Harvesting:** It is a system of catching rainwater where it falls. In rooftop harvesting, the roof becomes the catchment, and the rainwater is collected from the roof of the house/building. It can either be stored in a tank or diverted to an artificial recharge system. This method is less expensive and very useful and, if implemented properly, helps in augmenting the groundwater level of the area

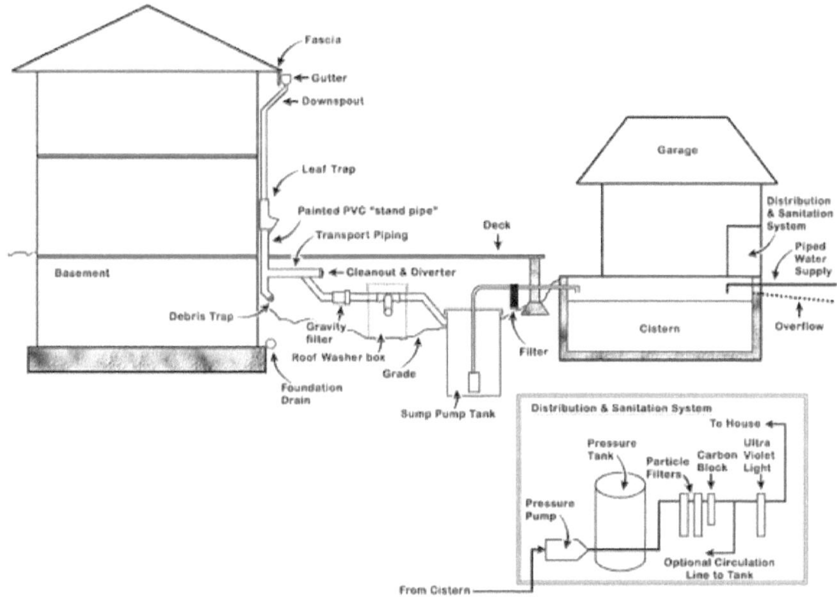


**Figure 1: A schematic representation of the Rainwater Harvesting System**

## ROOFTOP RAINWATER HARVESTING SYSTEM

Rooftop harvesting is the most popular way among the urban population. Some basic components of rooftop rainwater harvesting are as follows –

**Catchment:** In the most basic form of this technology, rainwater is collected in simple vessels at the roof's edge. Variations on this basic approach include collecting rainwater in gutters that drain to the collection vessel through downward pipes constructed for this purpose and/or the diversion of rainwater from the gutters to containers for settling particulates before being conveyed to the storage container for domestic use.

**Transportation:** Rainwater from the rooftop should be carried through down take water pipes or drains to the storage/harvesting system. Water pipes should be UV resistant (ISI HDPE/PVC pipes) of the required capacity. Water from sloping roofs could be caught through gutters and down take the pipe. At terraces, the mouth of each drain should have wire mesh to restrict the floating material.

**First Flush:** A first flush device is a valve that ensures that runoff from the first spell of rain is flushed out and does not enter the system. This needs to be done since the first spell of rain carries a considerably high quantity of pollutants from the air and the catchment surface.

**Filter:** This filter is used to remove suspended pollutants from rainwater collected from the roof. A filter unit is a chamber filled with filtering media such as fiber, coarse sand, and layers of gravel. It removes debris and dirt from the water before it enters the storage tank or recharge structure.

**Figure 2: A barrel filter in a rooftop rainwater harvesting system**

## Rooftop Rainwater Harvesting Potential

Rooftop rainwater harvesting Potential can be estimated by using Gould and Nissen Formula (1999):

**S = R. A. Cr**

Where S – Potential of roof rainwater harvesting (in $m^3$)

R = Average annual rainfall in m.

A = Roof area in $m^2$.

$C_r$ = Coefficient of Runoff

The runoff (Cr) coefficient for any catchment is defined as "The ratio between volumes of water that runs of and that of the total volume of rain that falls on the Rooftop." The Asbestos roof is 0.85, for the Concrete roof, it is 0.95, and for the GI roof, it is 0.98.

## Methods of Rooftop Rainwater Harvesting

### *Storage of Direct use*

In this method, rainwater collected from the roof of the building is diverted to a storage tank. The storage tank must be designed according to water requirements, rainfall, and catchment availability. Each drainpipe should have a mesh filter at the mouth and first flush device followed by a filtration system before connecting to the storage tank. Each tank should have an excess water overflow system.

Excess water could be diverted to the recharge system. Water from the storage tank can be used for secondary purposes such as washing and gardening etc. This is the most cost-effective way of rainwater harvesting. The main advantage of collecting and using rainwater during the rainy season is saving water from conventional sources and saving energy incurred on transportation and water distribution at the doorstep. This also conserves groundwater if it is being extracted to meet the demand when rains are on.

### Recharging groundwater aquifers

Various structures can recharge groundwater aquifers to ensure the percolation of rainwater in the ground instead of draining it away from the surface. Commonly used recharging methods are: -

a) Recharging of bore wells

b) Recharging of dug wells.

c) Recharge pits

d) Recharge Trenches

e) Soak ways or Recharge Shafts

f) Percolation Tanks

## OPERATION AND MAINTENANCE

The operation and maintenance of a system is the continuous process of checking to see if individual system components function properly, observe storage volume, and monitor water usage. Routine maintenance and proper upkeep are directly related to water quality for potable water systems —incorrect or deficient equipment results in lower water quality and increased health risks.

### Procedures

1) Become familiar with all the maintenance procedures, and double-check all devices and components before retaining any water.

2) Inspect and clean gutters, downspouts, conveyance piping, screens, and all devices upstream of the tank.

3) Determine through flushing that all components upstream of the tank are working properly.

4) In some cases, divert 100 percent of the rainwater from the tank until

the conveyance piping is no longer flushing construction residue or the chlorine solution used to clean the system.

5) If possible, fill the tank with potable water from a reputable source.

6) All faucets, outlets, and piping with non-potable water should be labelled.

7) Check for leaks.

8) Add pressure to the lines slowly by opening downstream valves to allow air to escape.

9) Open potable water faucets to purge non-treated water.

10) Test water quality parameters: a. Chlorine residual, b. Bacteria – TC/FC, Giardia, Cryptosporidium, Turbidity, and pH, etc.

Some of these operation and maintenance procedures are discussed in details –

## Gutters

Gutters are designed to catch all the runoff water from a roof. This clever but simple design also results in trapping debris and eventually blocking the flow of water. Monthly inspections of the gutter and removal of all materials, mostly organic matter, is necessary to maximize water quality. Additionally, the gutters should be inspected after high-intensity storms that include powerful wind gusts. At least once a year, gutters should be flushed to remove sediment and debris lodged in corners, transitions, and internal hangers.

**Figure 3: Debris lodged in a gutter can block the downspout screen and result in a flooded gutter.**

## Roof Washers and First Flush Diverters

Roof washers, some box filters, and first flush diverters are considered a second defence line against contamination after screening debris. Like a gutter, blockage in this device has negative consequences that result in less than optimal system performance and water quality. These devices are natural traps for sediment and organic matter; a weekly inspection is necessary. Monthly cleaning is suggested, depending on the volume of debris encountered. The drains should be kept clear to prevent puddling of water.

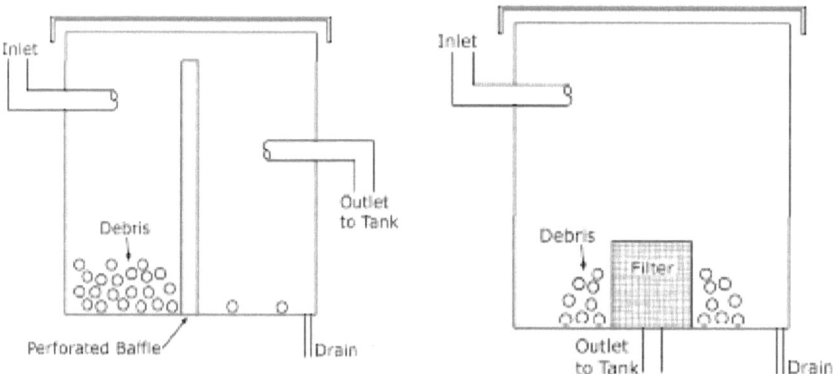


**Figure 4: Roof washers can become plugged with debris, and its removal is necessary to maximize water quality.**

## Piping and Connections

Piping and connections should be checked on a monthly basis. Plastic pipes should be checked for cracks and deformation. PVC plastic that is exposed to sunlight can degrade, turn yellowish, and become brittle. A visual inspection of outdoor components and piping should be conducted for any abrupt change of temperature. Repairs that involve replacement and reconnecting of system components should be inspected more often until it is determined that there are no leaks.

## Filters

As the filter's surface becomes clogged with particles, the flow is hampered, resulting in a drop in pressure results. A water pressure gauge installed on both the upstream and downstream sides of a filter or bank of filters can indicate a pressure drop. This shows the required maintenance. Some filters can be cleaned while others, mainly charcoal, must be replaced. Charcoal filters are replaced after a certain quantity of water has passed through them.

## Water Testing

Before consuming the water, an initial water quality assessment should be completed. An individual should evaluate with adequate knowledge and experience. Baseline test results provide a benchmark for compare subsequent results. At least, the water should be tested for bacteria, cryptosporidium, and giardia. The original analysis should be kept on file. The system should be retested after major repairs or replacement of sanitation equipment. If an unexpected or unexplained change in water quality occurs, testing for contamination may be appropriate. Yearly testing for total coliform (TC) and fecal coliform (FC) should be completed to indicate that the system is continuing to work properly. A client may view testing as expensive and unnecessary, but it ensures that the water that is being delivered remains at an acceptable quality.

## SURFACE RUNOFF HARVESTING

Surface runoff water harvesting is the collection, accumulation, treatment, or purification, and storing of storm water for its eventual reuse. It can also include other catchment areas from manmade surfaces, such as roads, or different urban environments such as parks, gardens, and playing fields. Surface runoff water is an excellent alternative to using mains drinking water for many purposes. If properly designed, Surface runoff catchment systems can collect large quantities of rainwater. The main challenge Surface runoff water harvesting poses is removing pollutants to make this water available for reuse. Small reservoirs with earthen bunds or embankments to contain runoff or river flow are built from soil excavated from within the reservoir to increase storage capacity, and a spillway or weir allows controlled overflow when storage capacity is exceeded. The reservoirs can vary in size from less than a hectare to up to 12 ha.

## LEGISLATION AND MANDATES OF DIFFERENT STATES IN INDIA ON RAINWATER HARVESTING

### A. *NTPC*

NTPC has released a guideline on Rain Water Harvesting in the Year 2018. In the policy document, it is written:

*"To strengthen its water conservation initiatives, NTPC has developed Rain Water Harvesting Policy 2018 which would act as the major guiding document for rain water harvesting. This policy is integral to NTPC Water Policy 2017 and shall be considered as its extension. Considering the importance of water as a shared resource, this policy can be further used as a reference for the various water conservation initiatives taken up by NTPC under CSR/ SD projects.*

### *Objective*

*To promote the installation and periodic upkeep of Rain Water Harvesting system in locations of and near NTPC establishments.*

### *Applicability & Scope*

*This policy is applicable to*

- *All the establishments of NTPC thus including projects, stations, administrative offices, residential townships, and guest houses;*

- *Locations with and without existing rainwater harvesting system;"*

### B. *Different Indian States*

#### 1. Himachal Pradesh

All commercial and institutional buildings, tourist and industrial complexes, hotels, etc., existing or coming up and having a plinth area of more than 1000 square meters will have rainwater storage facilities commensurate with the roof area's size. Unless the building owners comply with the new laws, they will not be issued 'No Objection' certificates required under different states. Toilet flush systems will have to be connected with the rainwater storage tank. It has been recommended that the buildings will have a rainwater storage facility commensurate with the size of the roof in the open and set back area of the plot at the rate of 0.24 ft$^3$/m$^2$ of the roof area.

## 2.  Ahmadabad

In 2002, the Ahmadabad Urban Development Authority (AUDA) had made rainwater harvesting mandatory for all buildings covering an area of over 1,500 square meters. According to the rule, for a coverage area of over 1,500 square meters, one percolation well is mandatory to ensure groundwater recharge. For every additional 4,000 square meters cover area, another well needs to be built.

## 3.  Bangalore

To conserve water and ensure groundwater recharge, the Karnataka government in February 2009 announced that buildings constructed in the city would have to adopt a rainwater harvesting facility compulsorily. According to the new law, residential sites, which exceed 2400 ft$^2$ (40 x 60 ft), shall create a rain harvesting facility.

## 4.  Port Blair

In 2007, Port Blair Municipal Council (PBMC) directed all the persons related to construction work to provide a proper spout or tank for collecting rainwater for various domestic purposes other than drinking. As per the existing building by-laws 1999, the building's slab or roof would have to be provided with a proper spout or gutter for rainwater collection, which would be beneficial for the municipal area residents during the water crisis.

The PBMC had advised all the owners of buildings in the Municipal area to comply with the provisions within four months, failing which action would be taken against them by the Council.

## 5. Chennai

Rainwater harvesting has been made mandatory in three-storied buildings (irrespective of the rooftop area's size). All new water and sewer connections are provided only after the installation of rainwater harvesting systems.

## 6. Kerala

The Kerala Municipality Building Rules, 1999 was amended by a notification dated January 12, 2004, issued by the Government of Kerala to include rainwater harvesting structures in new construction.

## 7. New Delhi

Since June 2001, the Ministry of Urban Affairs and Poverty Alleviation has made rainwater harvesting mandatory in all new buildings with a roof area of more than 100 sq. m. In all plots with an area of more than 1000 $m^2$, which are being developed.

## 8. CGWA

The Central Ground Water Authority (CGWA) has made rainwater harvesting mandatory in all institutions and residential colonies in notified areas (South and southwest Delhi and adjoining areas like Faridabad, Gurgaon Ghaziabad). This is also applicable to all the buildings in notified areas that have tube wells. The deadline for this was for March 31, 2002. The CGWA has also banned the drilling of tube wells in notified areas.

### 9. Indore (Madhya Pradesh)

Rainwater harvesting has been made mandatory in all new buildings with an area of 250 sq. m or more. A rebate of 6 percent on property tax has been offered as an incentive for implementing rainwater harvesting systems.

### 10. Kanpur (Uttar Pradesh)

Rainwater harvesting has been made mandatory in all new buildings with an area of 1000 sq. m or more.

### 11. Hyderabad (Andhra Pradesh)

Rainwater harvesting has been made mandatory in all new buildings with an area of 300 sq. m or more. The tentative for enforcing this deadline was June 2001.

### 12. Tamil Nadu

Through an ordinance titled Tamil Nadu Municipal Laws ordinance, 2003, dated July 19, 2003, the Government of Tamil Nadu has made rainwater harvesting mandatory for all the buildings, both public and private, in the state. The deadline to construct rainwater harvesting structures is August 31, 2003. The ordinance states, *"Where the rainwater harvesting structure is not provided as required, the Commissioner or any person authorized by him in this behalf may, after giving notice to the owner or occupier of the building, cause rainwater harvesting structure to be provided in such building and recover the cost of such provision along with the incidental expense thereof in the same manner as property tax."* It also warns the citizens on disconnection of water supply connection provided rainwater harvesting structures are not provided.

### 13. Haryana

Haryana Urban Development Authority (HUDA) has made rainwater harvesting mandatory in all new buildings irrespective of roof area.

In the notified areas in Gurgaon town and the adjoining industrial sites, all the institutions and residential colonies have been asked to adopt water harvesting by the CGWA. This is also applicable to all the buildings in notified areas having a tube well; the deadline was March 31, 2002. The CGWA has also banned the drilling of tube wells in notified areas.

### 14. Rajasthan

The state government has made rainwater harvesting mandatory for all public and establishments and all properties in plots covering more than 500 sq. m in urban areas.

### 15. Mumbai

The state government has made rainwater harvesting mandatory for all buildings constructed on plots that are more than 1,000 sq. m in size. The deadline set for this was October 2002.

### 16. Gujarat

The state roads and buildings department has made rainwater harvesting mandatory for all government buildings.

## CONCLUSION

Harvesting and collecting rainwater is an adequate strategy that can be used to address the water crisis problem globally. The use of a rainwater harvesting system provides excellent merits for every community. This simple water conservation method can boost an excellent solution in areas where there is enough rainfall but not enough groundwater supply. It will provide the most sustainable and efficient means of water management and unlock the vista of several other economic activities, leading to people's empowerment at the grass-root level.

Major cities like Delhi, Mumbai, Bangalore, Hyderabad, as well as tier 2 cities such as Chandigarh, Indore, Surat, and Nagpur, do have laws regarding rainwater harvesting. However, it is not good enough for such rules existing just on paper. If the concerned authorities fail to check to regularly ensure their on-ground implementation, rainwater harvesting rules are toothless.

For this, the Government should come out with an appropriate incentive structure and logistic assistance to make it a real success. Rainwater harvesting is something that thousands of families worldwide should participate in rather than pinning hopes on the administration to fight the water crisis. This water conservation method is a simple and effective process with numerous benefits that can be easily practiced in individual homes, apartments, parks, and across the world. As we all know, charity begins at home; likewise, a contribution to society's welfare must be initiated from one's home.

# 5

## SMART LEAK DETECTION TECHNOLOGIES

Sabarna Roy[1]

1. *Business Development, Applications Technology, Engineering and Strategy Department, Senior Vice President, Electrosteel Group, Kolkata-700017, India*

Kaustav Ray Chaudhury[2]

2. *Business Development, Applications Technology, Engineering and Strategy Department, Senior Executive, Electrosteel Group, Kolkata-700017, India*

Pipelines are widely used for the transportation of water over millions of miles all over the world. The structures of the pipelines are designed to withstand several environmental loading conditions to ensure safe and reliable distribution from point of production to the shore or distribution depot. However, leaks in pipeline networks are one of the major causes of innumerable losses in pipeline operators and nature. Incidents of pipeline failure can result in serious ecological disasters, human casualties and financial loss. In order to avoid such menace and maintain safe and reliable pipeline infrastructure, substantial research efforts have been devoted to implementing pipeline leak detection and localisation using different approaches. This paper discusses pipeline leakage detection technologies and summarises the state-of-the-art achievements.

The major causes of pipelines failure are pipeline corrosion, human negligence, defects during the process of installation and erection work, and flaws occurring during the manufacturing process and external factors.

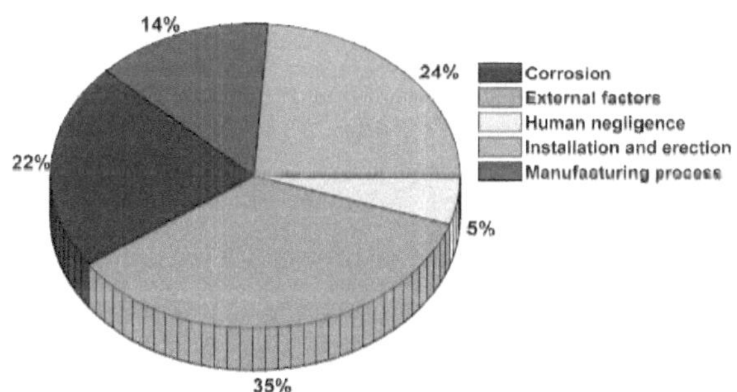

## Pipeline Leakage Detection Methods

1. **Exterior methods**

   - *Acoustic sensing*
   - *Accelerometer*
   - *Fibre optic sensing*
   - *Vapour sampling*
   - *Infrared thermography*
   - *Ground penetration*
   - *Fluorescence*
   - *Electromechanical impedance*
   - *Capacitive sensing*

2. **Interior/computational based methods**

   - *Mass/Volume*
   - *Negative pressure*
   - *Pressure point analysis*
   - *Digital signal*
   - *Dynamic modelling*
   - *State estimators*

## Exterior Based Leak Detection Methods

Exterior methods mainly involve the use of specific sensing devices to monitor the external part of the pipelines. The operational principle, strengths and weaknesses of these methods are discussed in the subsequent sections.

1. **Acoustic Emission Sensors**

When a pipeline leak occurs, it generates elastic waves in the frequency range up to 1 MHz due to high-pressure fluid escaping from the perforated point that allows one to detect pipeline leakage incidents. The time lag between the acoustic signals sensed by two sensors is employed to identify the leakage position. Acoustic methods for leak detection can be divided into two classes: active and passive. Active methods detect pipeline defects by listening to the reflected echoes of sound pulses emitted due to leakage. On the contrary, passive methods detect defects by listening to changes in sound generated by pressure waves in the pipelines.

2. **Accelerometers**

Accelerometers are another type of vibro-acoustic measuring device that is also useful to monitor low-frequency pipe-shell vibrations. The use of both accelerometers and hydrophones for monitoring pipelines was proposed.

3. **Fibre Optic Method**

This method involves the installation of fibre optic sensors along the exterior of the pipeline. The sensors can be installed as a distributed or point sensor to extensively detect the variety of physical and chemical properties of hydrocarbon spillage along the pipelines. The operation principle of this method is that cable temperature will change when pipeline leakage occurs and hydrocarbon fluid engross into the coating cable. By measuring the temperature variations in fibre optic cable anomalies along the pipeline can be detected

4. **Vapour Sampling Method**

Vapour sampling is generally used to determine the degree of hydrocarbon vapour in the pipeline environment. Though it is applicable in gas storage tank systems, it is also suitable to determine gas discharges into the environment surrounding the pipeline.

5. **Infrared Thermography**

Pipeline leakage detection systems based on the infrared thermography (IRT) mechanism are also applicable for the detection of pipelines leakages. IRT is an infrared image-based technique that can detect temperature changes in the pipeline environment using infrared cameras which shows the infrared range of 900–1400 nm. The image captured using an IR thermography camera is referred to as a *thermogram*.

6. **Ground Penetration Radar**

The emergence of ground penetration radar (GPR) is considered as an environmental tool that is valuable to detect and identify physical structures such as buried pipelines, water concentrations and landfill debris in the ground. GPR is a non-invasive high-resolution instrument that utilises electromagnetic wave propagation and scattering techniques to detect alterations in the magnetic and electrical properties of soil in the pipeline surrounding.

7. **Fluorescence Method**

Fluorescence methods for hydrocarbon spill detection employ light sources of a specific wavelength for molecule excitation in the targeted substance to a higher energy level. The detection of the spill is based upon the proportionality between the amount of hydrocarbon fluid discharged and the rate of light emitted at a different wavelength which can then be picked up for detection of the occurrence of hydrocarbon spillage.

8. **Capacitive Sensing**

In this technique, the change in the dielectric constant of the medium surrounding the sensor is measured to identify the existence of hydrocarbon spillage. The capacitive sensor is a local coverage point sensor that is generally employed in subsea pipelines. Capacitive sensor has been introduced to the market for environmental monitoring. Besides, buoyancy effects may carry the leaking medium away from the sensor vicinity which can be overcome by installing a collector for hydrocarbon spills over the monitoring structure.

## 9. Electromechanical Impedance-Based Methods

In electromechanical impedance (EMI)-based techniques a variation in structural mechanical impedance instigated by the incidence of pipeline damage is monitored to detect the occurrence of pipeline failure. In the event of pipeline defects, the EMI employs high-frequency structure excitation (usually greater than 30 kHz) through a surface-bounded piezoelectric sensor to sense variations in structural point impedance.

## Interior/Computational Methods

Interior or computational methods utilise internal fluid measurement instruments to monitor parameters associated with fluid flow in pipelines. Details of each of these techniques are discussed in the subsequent sections.

## 1. Mass-Volume Balance

The mass-volume balance approach for leak detection is straightforward. Its operation is based on the principle of mass conservation. The principle states that a fluid that enters the pipe section remains inside the pipe until it exits from the pipeline section. In a normal cylindrical pipeline network, the inflow and outflow fluid can be metered. In the absence of leakage, the assumption is that the inflow and outflow measured at the two ends of the pipeline section must be balanced, so a discrepancy between the measured mass-volume flows at the two ends of the pipeline indicates the presence of leakage.

2. **Negative Pressure Wave**

Leak detection techniques using negative pressure waves (NPWs) are based on the principle that when leakage occurs, it causes a pressure alteration as well as a decrease in flow speed which results in an instantaneous pressure drop and speed variation along the pipeline. As the instantaneous pressure drop occurs, it generates a negative pressure wave at the leak position and propagates the wave with a certain speed towards the upstream and downstream ends of the pipe. The wave contains leakage information which can be estimated through visual inspection and signal analysis.

3. **Pressure Point Analysis**

The pressure point analysis (PPA) method is a leak detection technique based upon the statistical properties of measured pressures at different points along the pipeline. The leakage is determined through the comparison of the measured values against the running statistical trend of the previous measurements.

4. **Digital Signal Processing**

In digital signal processing approaches, the extracted information such as amplitudes, wavelet transform coefficients and others frequency response is employed to determine leakage events.

5. **Dynamic Modelling**

Dynamic modelling-based pipeline leak systems are gaining considerable attention as they appear to be a promising technique for the detection of anomalies in both surface and subsea pipeline networks. In this approach, mathematical models are formulated to represent the operation of a pipeline system based on physics principles. The detection of leakages using this method is performed from two different points of views: (1) a statistical point of view and (2) a transient point of view.

6. **State Estimators/Observers Method**

State estimator or observer is a method that is based on dynamical modelling of flow process in the pipeline to estimate or observe variations in the variables associated with the fluid flow and indicate the occurrence of fault as a result of pipe damage.

## Research Gaps and Open Issues

Variations in physical parameters of the pipeline operation such as vibration, temperature, pressure etc. are expected to be detectable and communicated to reveal the incidences of anomalies. Leaks can only be accurately detected if the incident is within the vicinity of the monitoring sensor and thus the accuracy of leak detection systems becomes questionable if the leaks are not within the receptive fields of the sensors. However, the development of simple but realistic models for analysis and optimization remains challenging research questions. Since a high percentage of pipeline systems are made up of underground and underwater pipelines networks and the power required for real-time sensing and data communications in such environments is demanding, better replacement of sensor nodes in these settings is expensive or infeasible for large sensor networks. In order to achieve long-lived networks in these energy-constrained environments, different energy consumption minimization methods such as low energy adaptive clustering hierarchy, in-network processing, and sleep mode configuration have to be applied.

## Conclusions

This paper provides a reference in selecting an appropriate leak detection technology for a particular setting. A comprehensive study of various available pipeline leakage detection and localisation methods is discussed. Based on the study, it can be concluded that each technique has some merits and drawbacks. Most of the interior methods are sensitive to small leakages, especially if the point of leakage is close to the sensing device, but they are more prone to false alarms as they can easily be affected by environmental noise. Finally, after discussion on research gaps and open issues in pipeline leakage detection, characterization and localization we observed that despite having invested a considerable amount of research effort in pipeline leak detection and localisation systems, various gaps must still be filled before reliable real-time leakage detection in pipelines can be achieved fully.

# 6

## SMART WATER RECLAMATION & REUSE

Sabarna Roy[1]

1.  *Business Development, Applications Technology, Engineering and Strategy Department, Senior Vice President, Electrosteel Group, Kolkata-700017, India*

Kaustav Ray Chaudhury[2]

2.  *Business Development, Applications Technology, Engineering and Strategy Department, Senior Executive, Electrosteel Group, Kolkata-700017, India*

## INTRODUCTION

Water demand is increasing at an alarming rate, so the need for breakthrough technologies is required for sustainability and change of prevailing practices. The driving forces for change of prevailing practice point to demographics, economic and societal situation, and moving from collective to individual. Efficiency and adaptability are the two foremost things that will pave the way for restoring and improving urban infrastructure and provide access to clean water. Smart Water Reclamation and Re-use are the ultimate solutions to counter the water scarcity in the world.

## WATER RE-USE TECHNOLOGIES

Water Re-use and its use in non-potable applications can only be achieved through advanced technologies such as membrane technology and advanced oxidation process. The Advanced Water Treatment Systems (AWTSs) for water reclamation are:

**Preliminary Treatment:**

*   To remove the floating & suspended solids matter.

**Advanced Treatment:**

- Reverse Osmosis (RO).

- Electro-dialysis (ED).

- Nano-filtration (NF).

- Ion-Exchange (IE).

- Advanced-adsorption systems (AC or modified-clay).

- Combinatory oxidation and filtration system(s).

**Advanced Oxidation Processes (AOPs):**

- Ozone; Ozone/hydrogen-peroxide; Ozone/bio filtration.

- UV/Chlorine; UV/hydrogen-peroxide.

Problems of Advanced Water Treatment Technologies are operational cost, sustainability and general feasibility. Nevertheless, technology transfer at the global level should be done for the advancement in breakthrough technologies for water reuse.

## WATER RE-USE POLICIES IN INDIA

The consequences of pollution by the wastewater and degradation of water sources is our big challenge. The natural purification process of water in rivers and water bodies suffers from inadequacy and "Stream Sanitation" as a natural process has failed due to pollution load. The steps taken by the government for water reclamation are as follows:

***Rajasthan - TWWR Policy, 2016:*** The policy lays down the guidelines for achieving 100% sanitation, wastewater as a resource and treated wastewater be used for agriculture and industries with due safeguards.

***Maharashtra - TWWR by Power Plants and Industrial Estates - Nov 2017:*** The Govt. of Maharashtra gives a deadline of 3 years. Thermal power plants and industries in MIDC will not get fresh water after three years. To achieve the policy objectives and goals, a time-bound schedule has been laid down.

***Gujarat - TWWR Policy, 2018:*** The Govt. of Gujarat Policy - May 2018 states to maximize the collection and treatment of the sewage and re-use the same on a sustainable basis to reduce the stress on freshwater sources. The policy considers wastewater as an economic resource. A time-bound schedule for full use of reclaimed water by 2030.

## ENABLING CIRCULAR ECONOMY (CE) IN URBAN WATER

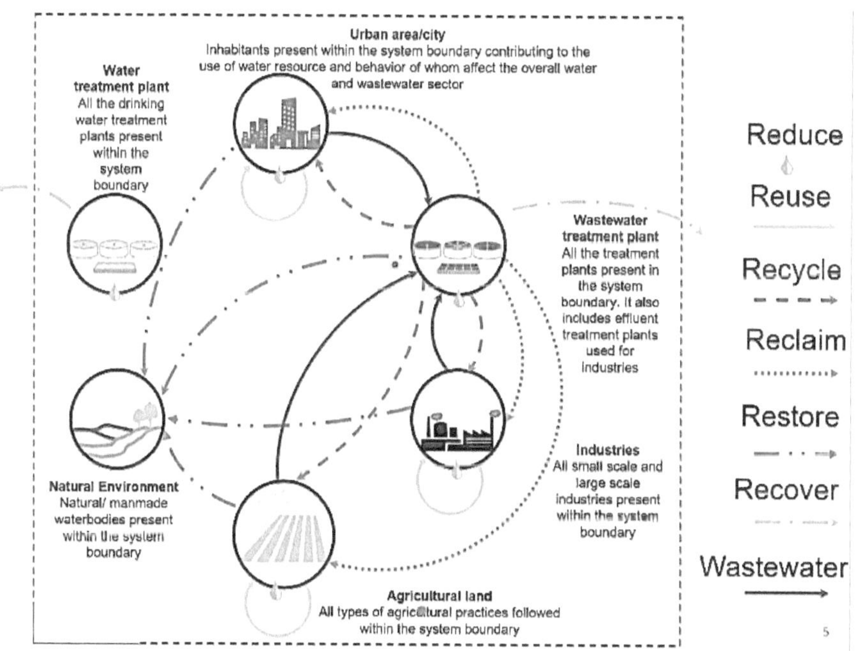


*Fig1: 6R Strategies for Circular Economy in Water Sector*

| | Systems thinking | Value Optimization | Innovation | Collaboration | Stewardship | Transparency |
|---|---|---|---|---|---|---|
| **Reduce** | | Optimal utilization of freshwater by using water-efficient fixtures at the household level enhances the overall availability of freshwater. reducing wastewater generation. | | Collaboration between consumers, suppliers, and government to provide innovative practices (e.g. nudges) can reduce the water demand and enhance the value of water, which was otherwise ignored. | Monitoring reduction at the source level, thus assessing direct and indirect impacts of reducing consumption with an authentic and transparent exchange of information internally and externally. | |
| **Reuse** | | | | Collaboration between water consumer and water utilities to design the reuse activity considering sustainable management of resources. For example, reusing water within the industry for cooling purpose enhances the value of used water. | Assessing positive and negative impacts of wastewater reuse in its crude form and hence maintaining transparency among authorities about the characteristics of wastewater and internally as well as externally. | |
| **Recycle** | | Cascading/reuse, recycling, and/or treatment/reclamation at household or industry level enhance the utility and value of used water thus significantly reducing the overall freshwater demand. | | Collaboration between WWTP, government, and consumer to design efficient and innovative recycling facility to enhance value creation. | | |
| **Reclaim** | | | | | | |
| **Recover** | | Wastewater is a potential resource for water recovery, energy, and material thus, is an opportunity to keep the waste (resources) present in wastewater at its highest utility and value, eventually reducing the extraction of fresh resources. | | Collaboration between waste management authorities, market and consumer to design reclamation and/or recovery facilities to generate/recover products as per the market value and consumer requirement considering the sustainability aspect. | Monitoring recycles, reclamation and recovery activities including direct and indirect impacts from economic, environmental and social aspects. | Transparency to exchange information about the methodology and extent of recycle, reclamation and recovery activities practised. |
| **Restore** | | Utilizing rainwater or treated wastewater for groundwater restoration can enhance the groundwater level reducing water scarcity in the dry season. Another example is afforestation that helps in restoring the water cycle, thus increasing the value of land as well as reducing the environmental pollution. | | Collaboration in between government authorities and customer to devise innovative groundwater restoration model to enhance the availability of freshwater in future e.g. rainwater harvesting. | Responsible action towards the restoration activities at small scale as well as large scale for promoting water restoration in the urban water cycle. | Transparency to share Information about the activities involved in the restoration and the extent of restoration achieved on a timely basis. |

***Table 1: Confluence of 6Rs and BS 8001:2017 Principles for Circular Economy in Water.***

## **CONCLUSIONS**

Considering all our traditional water sources, existing WWTPs constitute our captive and available water sources for treatment to the desired extent and reclaim water for re-use. The re-use of all treated wastewater for all human needs is imminent. The treatment plants must be re-designated as secondary water sources and termed Reclamation Plants.

The recycling market can be created by working with various stakeholders (and possible consumers) such as institutions, commercial establishments, industrial clusters, railways and metro boards, port trusts, the construction sector, premium consumers.

The Jal Shakti Ministry and water regulators should give guidelines to rename all STPs / WWTPs as SRPs / WWRPs while providing a conducive framework for treated wastewater re-use.

# 7

# SEAWATER DESALINATION, A RELIABLE AND SUSTAINABLE SOLUTION

Sabarna Roy[1]

1. *Business Development, Applications Technology, Engineering and Strategy Department, Senior Vice President, Electrosteel Group, Kolkata-700017, India*

Kaustav Ray Chaudhury[2]

2. *Business Development, Applications Technology, Engineering and Strategy Department, Senior Executive, Electrosteel Group, Kolkata-700017, India*

**Abstract:** The objective of the technical paper is to discuss the following objects in connection with Alleviating Domestic and Industrial Water Scarcity and advances in Membrane Technologies to Drive a Circular Economy.

- Need for Desalination- Driving factors.

- Present Major Desalination Technologies.

- Where does India stand in Desalination.

- Future Desalination Technologies under development.

- Factors influencing Desalinated water cost.

- The myth about Desalination and Industries source (raw water) for desalinated water.

- Innovations in Membrane Technology to enhance environmental friendly technology.

- Schematic diagram of a model Waste Water Treatment Process using Membrane Technology and how MLD provides affordability to ZLD for Textile Water.

- Various MLD Solution for cost reduction and low energy operation.

**Key words:** MSF, MED, SWRO, FO, MD, CD, ZLD, MLD, Membrane Technology, TDS, COD, MBR, MABR

## 1. Introduction

Access to sufficient quantities of water for drinking, domestic use and commercial and industrial processes is critical to health and wellbeing. With the growth of the world population, the availability of freshwater decreases; but with the advancement of desalination technologies, seawater has now become an interesting water resource as an alternative to cope with freshwater shortage. This process can be applied wherever a reliable resource of water is needed. Today about 600 million people face water scarcity worldwide. In India also more than 200 million people don't have access to safe drinking water. On the other hand, 50% of the World's population lives in cities along the coastline and about 71% of the Earth's surface is water-covered. Due to this, it is the need of the hour to desalinate seawater. Desalination technology is so advanced and economical nowadays that it can produce the desired quality of water at an affordable price. Desalination technology is so advanced and economical nowadays that it can produce the desired quality of water at an affordable price.

With increasing demands placed on limited freshwater resources, we must move beyond a linear economy model where we "take, make and dispose" of raw materials, to a circular approach in which raw materials are recycled and reused. Transitioning to a circular economy can help businesses reduce the consumption of limited resources to levels that can help offset increasing demands.

## 2. Desalination Techniques

The most popular and commonly used desalination technologies include – Multi-Stage Flashing Desalination (MSF), Multi-Effect Desalination Process (MED), Sea Water Reverse Osmosis Process (SWRO). *Almost each of these technologies includes five basic components – Intake System, Pre-Treatment System, Main Desalination Process, Post-treatment System and Outfall System.*

## *2.1 MSF* – **Multi-Stage Flashing Desalination**

Multi-Stage Flashing Desalination (MSF) [Fig. 1] is common for dual-purpose plants (water and electricity) and commercially competitive for very large size plants. The technology is based on repeated flash evaporation in successive stages (typically 15-36 stages) at lower vapour pressure. A simple flow diagram of MSF process is shown below:

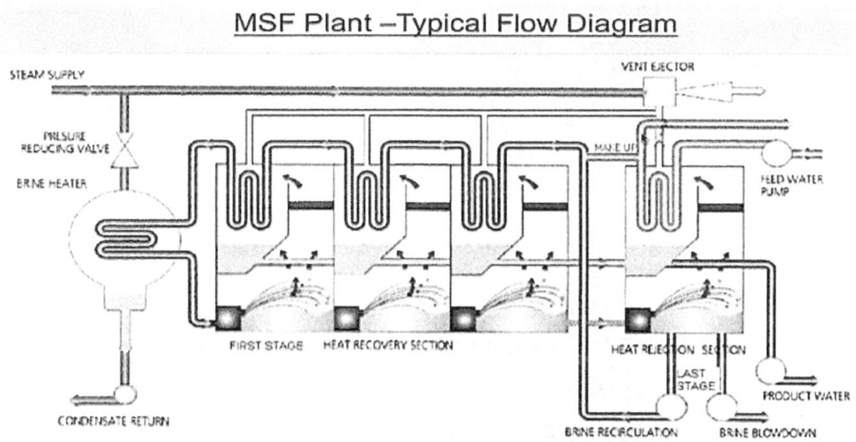


**Fig. 1 MSF Plant – Typical Flow Diagram.**

## *2.2 Multi-Effect Desalination Process (MED)*

Multi-Effect Desalination Process (MED) [Fig. 2] takes over MSF technology due to lost cost capital associated with higher efficiency. It is competitive for small and medium-sized plants with typically 6-8 stages. The typical MED-TVC process is shown in the flow diagram below:

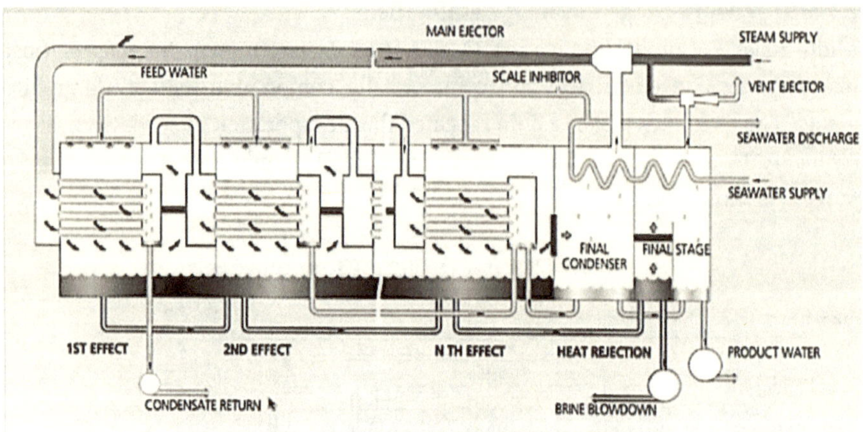


**Fig. 2 MED Plant – Typical Flow Diagram.**

### 2.3 *Sea Water Reverse Osmosis Process (SWRO)*

Sea Water Reverse Osmosis Process (SWRO) [Fig 3] is the most used cheapest desalination process for standalone units and commercially competitive for all sizes of plants and applications. It deploys high energy recovery devices with low pressure and very mature technology. SWRO process can be presented in the flow diagram below:

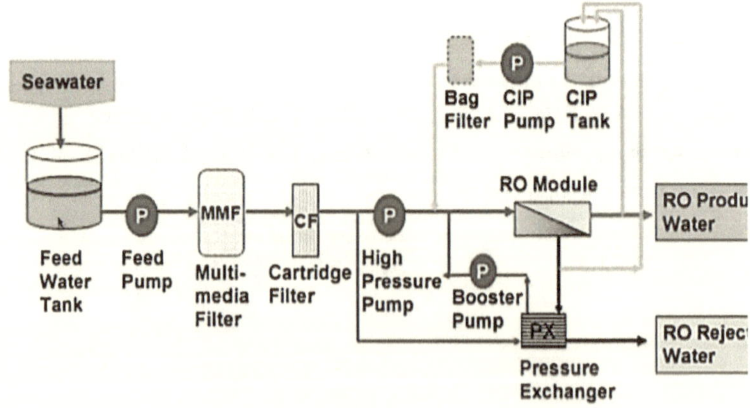


**Fig. 3 SWRO Plant – Typical Flow Diagram.**

In SWRO Plant the availability/reliability is achieved by the following points:

- Selection of proper seawater intake location.

- Extensive survey to decide the seawater quality for plant design.

- Designing the plant with a suitable pre-treatment system to supply the RO Train with consistent feed water quality meeting membrane manufacturer's requirement.

- Choosing proven equipment and membranes.

- Providing standby equipment.

- Choosing suitable MOC for equipment, tanks and piping to avoid corrosion.

- Adopting the right welding procedure for the High-Pressure Piping to avoid crevice/pitting corrosion.

- Predictive and preventive maintenance.

## 3. Desalination Plant Market Forecast

International Desalination Association (IDA) assessment [1] reveals that the total number of installed desalination plants worldwide, as of June 30, 2015, is 18,426, producing more than 86.8 million cubic meters per day. Global Water Intelligence (GWI) estimates the global water desalination market value at USD 13.31 billion in 2016.

According to "India Desalination Plant Market Forecast & Opportunities, 2019" [2], fig [4], the desalination plant market in India is projected to register a Compounded Annual Growth Rate of around 19% till 2019, growing at a faster rate than the global desalination market during the same period. As per 2017 statistics, India currently has around 500 desalination plants operating in different parts of the country.

**Indian Market Forecast 2013- 2020**

| Capital expenditure by sector ($m) | 2013 | 2014 | 2015 | 2016 | 2017 | 2018 | 2019 | 2020 |
|---|---|---|---|---|---|---|---|---|
| Utility water capital expenditure | 4,622.2 | 5,583.3 | 5,742.4 | 5,962.8 | 6,473.6 | 7,178.0 | 7,931.7 | 8,690.9 |
| Water networks | 2,425.6 | 2,937.5 | 3,093.8 | 3,238.8 | 3,523.1 | 3,858.6 | 4,222.5 | 4,620.8 |
| Water treatment plants | 661.5 | 801.1 | 843.8 | 883.3 | 960.9 | 1,052.3 | 1,151.6 | 1,260.2 |
| Water resources (excl. desalination) | 1,323.0 | 1,602.3 | 1,687.5 | 1,766.6 | 1,921.7 | 2,104.7 | 2,303.2 | 2,520.4 |
| Seawater and brackish water desalination | 212.1 | 242.4 | 117.3 | 74.0 | 67.9 | 162.3 | 254.3 | 289.4 |

Source: GWI 2017

**Fig. 4 Indian Market Forecast of Desalination Plants 2013 - 2020.**

## 4. Recent Developments in Desalination Technology

Innumerable developments are taking place in desalination technology. Selected important developments are –

1. Forward Osmosis (FO).

2. Membrane Distillation (MD).

3. Capacitive Deionization (CD).

4. Humidification-Dehumidification Process.

5. Solar Desalination.

6. Promising New RO Membrane Developments.

### 4.1 Forward Osmosis (FO)

Forward Osmosis (FO) [Fig. 5] uses a semi-permeable membrane to separate salts from water and it is an innovative membrane-based technology that has the potential to reduce the costs and environmental impacts of desalination. It uses an osmotic pressure gradient instead of hydraulic pressure, which is used in RO, to create the driving force for water transport through the membrane. No energy is needed to drive the water flux of an FO process, as the water flux is the natural tendency of the system. A simplified schematic process of FO using ammonium carbonate as a draw solution is given below:

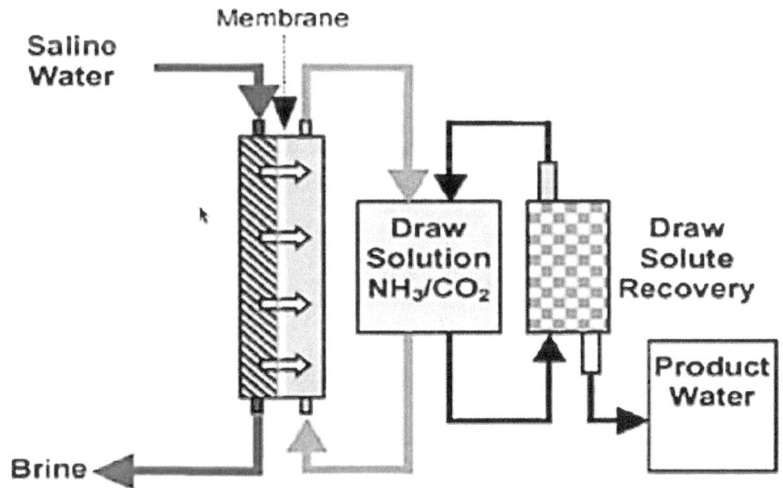


**Fig. 5 Schematic Process of Forward Osmosis (FO).**

### 4.2 Membrane Distillation (MD)

Membrane Distillation (MD) combines membrane technology and evaporation processing in a single unit. It functions at atmospheric pressure and requires only relatively low feed temperatures of 70°C to 90°C. MD transports water vapour through the pores of hydrophobic membranes using the temperature difference across the membrane and the membrane allows water vapour to penetrate the hydrophobic surface while repelling the liquid. The clean vapour is carried away from the membrane and condensed as pure water.

### 4.3 Capacitive Deionization (CD)

Capacitive Deionization (CD) is a low-pressure, non-membrane desalination technology that uses electrostatic forces to remove dissolved ions from the solution. Saline water is passed through pairs of electrodes held at a potential difference of 1.2 volts (V). The electrodes consist of porous carbon aerogel, with a high specific surface area and with very low electrical resistivity. Ions are adsorbed to the electrode of opposing charge in a semi-batch process. Eventually, the electrodes become saturated with ions and must be regenerated. To regenerate CD, the applied potential is removed, and the ions attached to the electrodes are released and flushed from the system. Flushing of cells using a small quantity of product water generates a concentrated stream. Unlike ion exchange processes, no additional chemicals are required for the regeneration of the electro sorbent in this system.

### 4.4 Humidification-dehumidification (HDH)

Humidification-dehumidification (HDH) [Fig. 6], desalination mimics the natural water cycle to desalinate the water. HDH desalination involves two processes – Seawater is first converted to water vapour by evaporation into dry air in an evaporator (humidification). This water vapour is then condensed from the air in a condenser to produce freshwater (dehumidification). Normal atmospheric air is used as the medium to convert seawater to freshwater. Heat for evaporation can be obtained from various sources, including solar, thermal, geothermal, and combinations of these. A flow diagram of HDH is shown below:

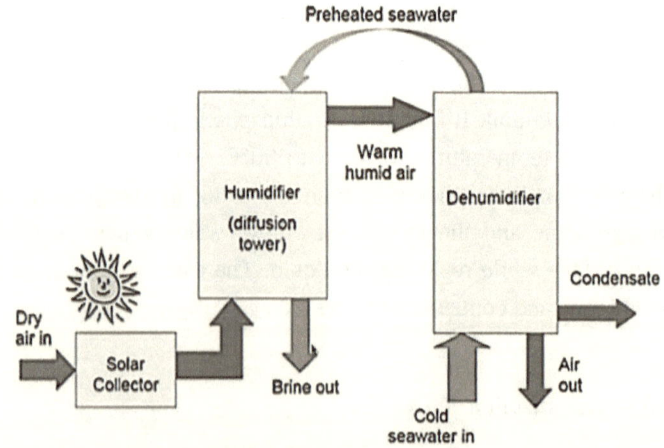


**Fig. 6 Flow Diagram of Humidification-dehumidification (HDH).**

### *4.4 Solar-dome desalination technology*

The solar-dome desalination technology uses concentrated solar energy by strategically arranged banks of mirrors that would speed up the natural process of evaporation within the domed greenhouse structures. The inflow of seawater, drawn by gravity, travels via enclosed glass aqueducts into the dome, where it is boiled, distilled and then collected in cauldrons. The dense steam produced falls as a tropical rainstorm within the dome as clean and pure water which is then piped to reservoirs in order to serve the local needs in the water. Some facts about costing in Solar Desalination with Solar Power:

- Marubeni had bid \$0.0157/kWh (INR 1.13/kWh) for an 800MW photovoltaic project in Qatar [3].

- Electricity contributes to >70% of SWRO Plant O&M cost in Nemmeli Desalination Plant as the electricity cost in Tamil Nadu is 6.35 INR/kWh [4].

- If the power cost goes down to INR 1.13 instead of INR 6.35 it will be an 82% reduction in power cost. However, we need such a breakthrough in electricity storage.

Some very promising New Membrane Developments are – a) Nano Engineered Membranes, b) Carbon Nanotube Membranes and c) Aquaporin Membranes.


## 5. Advances in Membrane Technologies to Drive a Circular Economy Results

Due to less availability of freshwater sources, manufacturing industries, especially textile industries, need alternative methods of optimizing their water usage. Moreover, stringent government regulations have limited the possibility of discharging the effluents to surface water bodies. Therefore, new technologies are sought after in wastewater treatment for the recovery and reuse of effluents. Hence, the tradition from Minimum Liquid Discharge (MLD) to Zero Liquid Discharge (ZLD) [Fig. 7] merges sustainability and innovation for a healthier future. Membrane technology is the emerging technology that will benefit industries and the environment alike.

The rising price of water and high discharge mitigation costs have prompted the search for an alternative to zero liquid discharge (ZLD). ZLD can be expensive and not necessarily environmentally friendly because of the energy and resources typically required to get discharges down to zero.

Some industrial and municipal users are turning to a minimal liquid discharge (MLD) approach: a core set of proven ultrafiltration, reverse osmosis, Nanofiltration, and ion exchange-based technologies and processes that enable users to achieve up to 95% liquid discharge recovery but at a fraction of ZLD's costs.

DuPont Water Solutions [5] can help address your wastewater regulatory challenges with technology that can minimize your operating costs and maximize water recovery while reducing the amount of energy required to operate.

The table given below illustrates how ZLD & MLD can be achieved through recovery processes with respect to Total Dissolved Solids (TDS) and Chemical Oxygen Demand (COD).

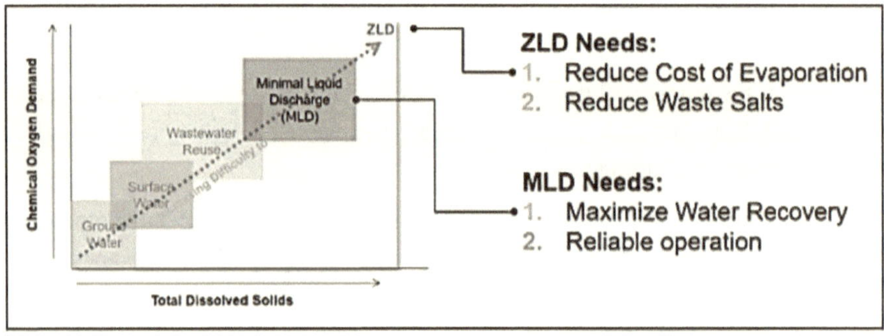


**Fig. 7 ZLD and MLD recovery process w.r.t TDS and COD.**

### 5.1 Cost of implementing Membrane Technology

It must be noted that ZLD is an extreme treatment process used to maximize water recovery, meeting stringent government regulations. But, the total treatment cost of ZLD is very high. So, Reverse Osmosis proves to be an effective way of reducing the cost. With ensuring higher recovery, reverse osmosis also helps in retrieving the brine solution, which further provides the dyes needed to textile industries.

The table given below shows that MLD/ZLD [Fig. 8] using RO water recovery provides CAPEX & OPEX savings.

|  | Est. Capital Costs $ per m³/d Feed | Est. Energy Cost kWh per m³ Feed |
| --- | --- | --- |
| Evaporator | 2000 – 6000 | 22 – 40* |
| RO 1 (low pressure) | 200 – 400 | 0.6 – 1 |
| RO 2 (high pressure) | 400 – 800 | 1.5 – 2 |
| RO 3 (ultra-high pressure) | 600 – 1200 | 5 – 8 |

**Fig. 8 Cost of Membrane Technology.**

*5.2 Impact of System Recovery on various parameters*

As the recovery percentage of the treatment system increases, the concentration of salts also increases [Fig.9]. From the graph shown below, it becomes quite clear that salt concentration increases exponentially once the treatment process aims from MLD to ZLD, thus making a recovery more difficult.

## Concentrate enrichment at high recovery

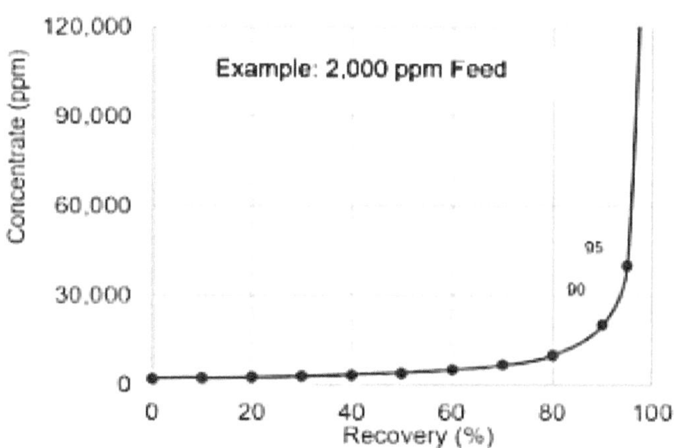


**Fig. 9 Concentration Vs Recovery.**

Therefore, specific challenges are to be overcome to increase recovery rates.

### *5.3 Innovations in Membrane Technology*

<u>Reverse Osmosis</u>

- Leading player with global scale
- Unique membrane chemistry and element design
- Dry sea & brackish water RO technology
- Quality S consistency

<u>Ion Exchange Technology</u>

- Market and Technology leader
- Broadest portfolio serving multiple markets
- Highest product quality and global footprint

<u>Ultra-Filtration</u>

- Acquisitions have positioned DWS as a market leader
- Broadest portfolios available in market Acquisition focus to broaden the offering to solve customer problems.

<u>Closed Circuit Reverse Osmosis (CCRO)</u>

- Breakthrough in RO technology with patent position
- Strong value proposition in core and attractive markets
- Significant improvements in recovery, brine concentration footprint, energy savings and self-operated.

<u>Membrane Bio-Reactor (MBR)</u>

- High growth product line-driven by regulations

- Opportunity for growth and market share increases

- Increased visibility on the advantage v competition a focus


## Membrane Aerated Bio-Film Reactor (MABR)

- New to the world technology

- Significant improvements in energy reduction, reduced footprint and OPEX savings.


## Membrane Contactors

- Virtual JV to bring DIC membrane contactor to water

- Strong need due to competitive situation and market growth

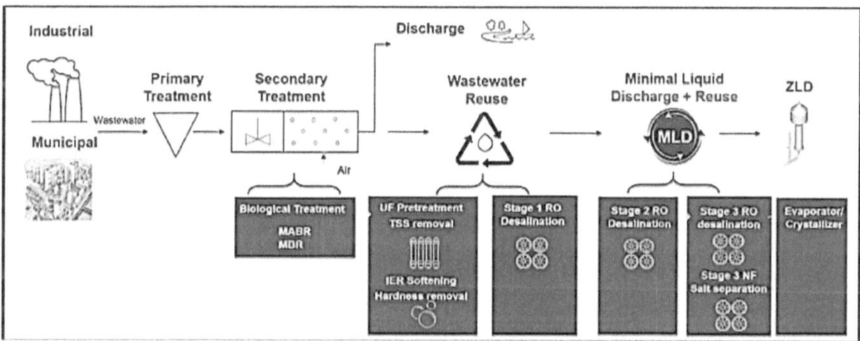


**Fig. 10 Schematic diagram of a model WW Treatment Process using Membrane Technology.**


## 6. Case Study on MLD and ZLD

Dupont Water Solutions organized a survey of four Textile Wastewater Treatment Site Operators [6]. For the OPEX analysis, the average treatment costs provided across the four plants were found to show Circular Economy Savings of $ 1.06/m³, the value of recycled water be $ 0.56/m³ and the amount of recycled salt generated as $ 0.50/m³. With the total cost of ZLD being $ 1.76/m³, the net OPEX of MLD/

ZLD came out to be $ 0.70/m³.

| Treat to discharge (if allowed) | | MLD/ZLD | | Thermal ZLD only | |
|---|---|---|---|---|---|
| Treatment cost | Discharge cost | Treatment cost | Recycle value | Treatment cost | Recycle value |
| $0.60/m³ | X | $1.76/m³ | $1.06/m³ | $3.48/m³ | $1.06/m³ |
| ~$0.60/m³ | | $0.70/m³ | | $2.42/m³ | |

**Fig. 10 Cost Comparison of MLD and ZLD.**

This case study proves that MLD brings affordability to ZLD for Textile Wastewater. Moreover, MLD/ZLD cost is similar to "treat to discharge" when considering the benefit of recycled water and salt value.

**7. Minimal Liquid Discharge Solutions and their Respective Advantages:**

INTEGRAFLUX™ XP ULTRAFILTRATION MODULES [5]

- High-productivity ultrafiltration modules with industry-leading membrane area and high permeability XP Fibre.

- High-mechanical-strength PVDF fibre with excellent chemical resistance providing long membrane life and reliable operation.

- Outside-in flow configuration allowing a wide range of solids in the feed water minimizing the need for pre-treatment processes and reducing the backwash volume compared to inside-out configurations.

FILMTEC™ FORTILIFE™ XC70 RO Elements [5]

- Maximize purity of reuse water while concentrating the brine to > 70,000 ppm TDS within standard operating limits.

- Fouling resistance.

- Increased productivity.

- Less frequent cleaning, more up-time, longer element life.

## FILMTEC™ FORTILIFE™ XC-N NF Elements [5]

- Highly selective membrane with high monovalent ion passage and divalent ion rejection.

- High permeability at low pressures and high TDS levels.

- Purified, high-concentration brine solution for reuse.

- Less dissolved solid wastes.

- Low-energy operation.

## DUPONT™ Specialty Membranes UHP RO Elements [5]

- Reduce brine volume and achieve concentrations of 100,000 – 200,000 ppm TDS, thus significantly reducing discharge volume and downstream processes.

- Distinctive ultra-high-pressure element construction allowing operation up to 120 bar (1,740 psi)

- High-pressure, fouling-resistant FILMTEC™ SW30 flat sheet.

## FILMTEC™ FORTILIFE™ XC80 RO Elements [5]

- Lower brine volume and maximize recovery by concentrating the brine to > 80,000 ppm TDS within standard operating limits.

- Fouling resistance.

- Reduce discharge with standard RO system designs.

- Less frequent cleaning, more up-time, longer element life.

- Low-energy RO operation.

## FILMTEC™ FORTILIFE™ CR100 RO Elements [5]

- Up to 50% less frequent cleanings due to biofouling.

- Improved hydraulic balance for low organic fouling and lower energy.

- Reduced rate of flux loss in challenging waters.
- Highly cleanable membrane chemistry.
- Up to 10% lower energy operation.
- High permeate quality to enable blending with higher TDS for reuse.

AMBERLITE™ IRC83 H WAC Resin for Hardness Removal [5]

- Up to 30% more operating capacity than current weak acid cation resins.
- Fewer regeneration cycles reduce waste volume up to 15%.
- Superior physical stability yields long resin life.

## 8. Conclusion

There are a lot of myths about Desalination Technology; it has a harmful effect on marine life, costs are exorbitant, desalinated water does not have minerals and it is therefore not good for health, and also desalination is a failure and the desalination plants are not reliable. Important facts/arguments against these myths are –

➢ Modern plant discharges are let deep into the sea through special diffuser arrangements in such a way that the discharge is diluted within a radius of 30m from the diffuser. After dilution, the salinity difference between brine discharge and seawater is < 5%.

➢ In the last 30 years, several inventions in energy recovery devices have reduced the power consumption from 8.5 kWh/m3 to less than 3.2 kWh/m3. International BOOT desalination projects Water Purchase Cost is in the range of 0.41 to 0.5 USD and BOOT water cost in a plant in Tamil Nadu is Rs. 56/ kWh (in 9th year) with electricity cost of Rs. 6.35 /kWh; with Solar power cost drastically falling the cost of desalinated water is bound to drop further.

➢ Seawater RO (SWRO) plants can be designed to give the required quality water. Generally, Municipal SWRO plants are designed to meet WHO and ISO standards. They produce water with Total Dissolved Solids in the range of 250 to 500 mg/l and it contains all minerals.

> The whole of the middle-east countries utility water demand is successfully met with desalinated water over decades. The availability of desalination plants is as high as 97% making them more reliable. SWRO plants can be started for a short time and also can be stopped as and when required.

Thus with increasing demands placed on limited freshwater resources, we must move beyond a linear economy model where we "take, make and dispose" of raw materials, to a circular approach in which raw materials are recycled and reused. Transitioning to a circular economy with the help of new technologies for desalination can help businesses reduce the consumption of limited resources to levels that can help offset increasing demands.

## 9. References


[1]   https://idadesal.org/.

[2]   www.globalwaterintel.com.

[3]   https://www.pv-magazine.com/2020/01/23/qatars-800-mw-pv-tender-saw-world-record-final-price-0-01567-kwh/

[4]   Design, Build 150mld Capacity Desalination Plant Based On Sea Water Reverse Osmosis At Nemmeli, East Coast Road, Chennai, Tamil Nadu And Operation And Maintenance For 20 Years Tender No: Cnt / WSS / ICB/Kfw/DESAL/012/2016-17.

[5]   https://www.dupont.com/water.html.

[6]   Textile Wastewater Treatment Options: A Critical Review, K. Siddique, M. Rizwan, M.J. Shahid, S. Ali, R. Ahmad, H. Rizvi.